W. F. Kirby

British Butterflies, Moths & Beetles

W. F. Kirby

British Butterflies, Moths & Beetles

ISBN/EAN: 9783337139742

Printed in Europe, USA, Canada, Australia, Japan

Cover: Foto ©berggeist007 / pixelio.de

More available books at **www.hansebooks.com**

Papilio Machaon.
(*Swallow-tail.*)

BRITISH
BUTTERFLIES, MOTHS
AND BEETLES.

BY

W. F. KIRBY,

*Of the Zoological Department, British Museum; Author of "An
Elementary Text Book of Entomology," "European
Butterflies and Moths," etc., etc.*

ARDVA · QVÆ · PVLCRA

LONDON:
W. SWAN SONNENSCHEIN & CO.,
PATERNOSTER SQUARE.
1885.

Butler & Tanner,
The Selwood Printing Works,
Frome, and London.

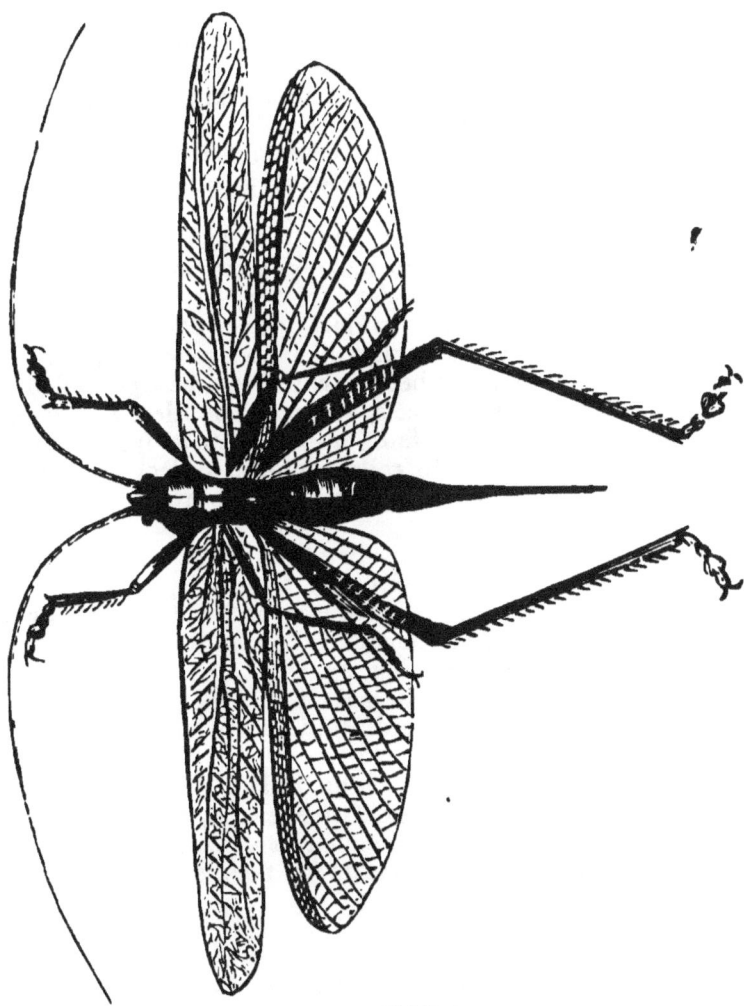

Phasgonura Viridissima.

[*Face p.* 3.

THE ORDERS OF INSECTS.

ENTOMOLOGY, or the Science of Insects, concerns a great number of living creatures, some of which we see around us every
day of our lives. They are far more numerous than any other
group of animals, for about 12,000 different kinds are known to
inhabit England, although the smaller and less attractive groups
are very insufficiently known at present ; and at least twenty
times this number are known to be found in other parts of the
world.

But in order to obtain a general knowledge of British insects,
it is by no means necessary to make yourself acquainted with
every one of these 12,000 species; for they have been divided
into sections, so that each individual species can be identified
and its resemblances to those most like itself perceived, and its
differences pointed out. When you have acquired a general idea
of the various sections of insects, you can then select the group
which you like best, and confine your attention to it ; but most
people, when they begin to collect insects, collect everything
which comes in their way, until they have formed this special
preference.

Naturalists have begun by dividing the various objects which
we see around us into the Animal, Vegetable, and Mineral
Kingdoms. The Animal Kingdom is again divided into several
large sections called Sub-Kingdoms, to one of which, called
variously *Arthropoda*, *Annulosa*, or *Articulata*, insects belong.
The *Arthropoda* have no internal framework of bones, like
vertebrate animals, but their bodies and limbs are formed of a
number of jointed pieces, of a bony or horny consistence, to
which the muscles are attached on the inside. This outer covering forms what is called their external skeleton, and its strength
and solidity is such that their activity and bodily powers are
frequently far greater in proportion to their size than in any
vertebrate animal.

The *Arthropoda* are again divided into four principal classes :
Crustacea, including Crabs, Lobsters, Shrimps, Wood-lice, etc. ;
Myriopoda, or Centipedes ; *Arachnida*, or Spiders, Ticks, and
Mites ; and *Insecta*, or Insects. We need not now discuss the

3

characters of the three first classes, as the Insects are separated from them by a great number of characters.

Insects have six legs in the perfect state, and no more ; four (or two) wings, two eyes, composed of a great number of facets, and sometimes one, two, or three eyes of another kind, called simple eyes, or stemmata, on the tip of the head. Their body is composed of thirteen segments, divided into head, thorax, and abdomen. They have neither heart, brain, nor nervous system at all resembling those of the higher animals. The place of the heart is supplied by an organ called the great dorsal vessel, lying along the back ; and the brain and nervous system of vertebrate animals are represented by a double row of connected ganglia, or knots of nervous matter, lying along the lower surface of the body. They breathe by means of spiracles, or air-holes, opening on each side of the greater number of the segments of the body. The muscular system is highly developed, the muscles being far more numerous than those of vertebrate animals. Insects pass through four stages, called respectively egg, larva (or caterpillar), pupa (or chrysalis), and imago, though these are more sharply defined in some insects than in others. They also moult their skins more or less frequently in the larva state ; and although they sometimes possess more than six legs in this state (and occasionally none at all), yet the larvæ of insects are not generally liable to be mistaken for any other animals.

We have spoken of the thirteen segments of which the body of an insect is composed ; the first forms the head, the second to the fourth the thorax, and the remainder the abdomen. These thirteen segments (except occasionally one or two of the terminal segments) are generally distinct in larvæ, but become more or less welded together in the perfect insect, in which, however, the three divisions of head, thorax, and abdomen are always distinctly visible, and are much more clearly defined than in the larva state.

The head contains the organs of sense, and the mouth. In addition to the eyes, there is always a pair of long jointed organs called antennæ, which appear to be organs of touch, smell, and probably of hearing. After the loss of these organs, an insect becomes wholly incapable of directing its flight. The antennæ differ very much in shape in different insects, and are called simple, pectinated, lamellated, clavate, etc., accordingly. Sometimes they are straight ; sometimes sharply angulated in the middle ; sometimes smooth ; sometimes hairy ; and often furnished with long projections, giving them the appearance of a comb, or of the feather of a bird.   Sometimes they are tapering at the end ; sometimes knobbed ; and sometimes, again, they have a series of long processes near the end, opening and closing at pleasure, almost like a fan.

The mouth of insects is formed either for biting or for suction. Those which have horny jaws are called mandibulate insects (*Insecta Mandibulata*), and those which are provided with a proboscis to imbibe liquid food are called haustellate insects (*Insecta Haustellata*). Most of the latter, however, are mandibulate in the larva state, and many mandibulate insects are likewise provided with a proboscis.

The three segments of the thorax are called the prothorax, mesothorax, and metathorax respectively.   The first bears on the under surface the first pair of legs.   The second bears the first pair of wings, and the second pair of legs ; and the third bears the second pair of wings, and the third pair of legs.   The under surface of the thorax is called the pectus, and the space beneath the wings the pleura.   The two pairs of wings are not always alike, and when there is any difference, the first pair are always thicker and narrower.   When they are much harder and thicker than the hind wings, so as to form wing-cases rather than additional organs of flight, they are called elytra.   The wings are always traversed by a greater or less number of jointed air-tubes, called nervures, the arrangement of which differs considerably in various insects.   The legs are divided into several parts.   First come the coxæ, or hips, which are generally the thickest parts of the leg ; next a connecting joint, called the trochanter ; after which follow two straight parts, called femora, or thighs, and tibiæ, or shanks, respectively.   Below these comes the foot, which is composed of five joints, called joints of the tarsi, and terminating in a pair of claws ; but in many insects, the claws, or even one or more of the joints of the tarsi themselves, are undeveloped.   The point of intersection of the femur and tibia is called the knee ; and the knees, like the trochanters, are occasionally of a different colour to the rest of the leg.   The legs, like the rest of the body, may be smooth, or clothed with hairs, or spines ; there are often a pair of long spurs at the end of the tibiæ, and sometimes also in the middle.

In many insects, the abdomen is completely covered by the wings when the insect is at rest ; while in other cases, it projects far beyond them.   Its latter extremity is often furnished with a variety of curious appendages, which are either directly or indirectly defensive or offensive weapons, or connected with the reproduction of the species, and oviposition.

Insects differ very much in size ; the smallest insect known is
said to be a four-winged fly, the larva of which lives in the egg
of a parasite of a North American bee.   This little creature has
beautifully formed wings, each of which resembles a single
feather.   It measures one-ninetieth of an inch in length.   Although this is an American insect, yet we have several allied
species in this country, and need not despair of ultimately
meeting with a still smaller insect in England   On the other
hand, some of the great tropical moths and locusts measure a
foot across the wings ; but we must be contented to regard the
Death's Head Hawk-Moth, which sometimes measures nearly
six inches across the wings, as our largest British insect.

Many of our readers will perhaps think 12,000 a very large
number of different kinds of insects to be found in one country ;
but insects are by no means so abundant in England as in the
adjacent parts of the Continent, nor are they so destructive to
our crops.   Islands are always poorer in plants and animals
than continents ; besides, when forests are cleared, and marshes
are drained, numbers of insects are destroyed, and those which
are confined to such localities are very likely to become exter-
minated.   It is almost certain that a few centuries ago, when
England was covered with marsh and forest, many insects must
have been abundant which are now rarely or never met with.
In fact, several different kinds are known to have become extinct
in England within the memory of many entomologists now
living ; and if this is the case among large and conspicuous
insects, it must also have happened to many small and incon-
spicuous kinds without our being even aware of it.   The south-
eastern counties having the finest and driest climate in the
islands, as well as being those nearest the Continent, produce
most species of insects.   Insects are much less numerous in
Scotland and Ireland than in England, though these parts of the
kingdom produce some species not to be met with elsewhere in
the British Islands.   Although nearly all our British insects are
at least as common on the Continent as with us (and often much
more so), yet there are a few species and varieties, generally
confined to very restricted localities, which have hitherto only
been met with on this side the water.

These 12,000 different kinds of insects are divided into seven
large sections, called Orders.   Some writers admit more, but
the seven great Orders are those which are universally recognised, and the smaller ones are now generally treated as forming
part of the others.   These date from the time of Linnæus, who
founded the modern system of classification, and are called

Calepteryx Virgo.

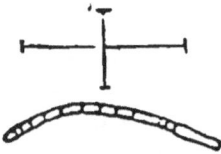

Apis Mellifica.

Smicra Sispes.

Panorpa Communis.

[Face p. 6.

Coleoptera, Orthoptera, Neuroptera, Hymenoptera, Lepidoptera, Hemiptera, and *Diptera ;* according to the general character of the wings in each Order. The first four Orders are mandibulate, and the three latter haustellate. But we must here point out that it is impossible to place insects in a linear arrangement which shall also be natural. Thus, even in the arrangement of the Orders, the *Hemiptera* might be placed between the *Orthoptera* and *Diptera ;* or the *Lepidoptera* between the *Neuroptera* and *Diptera,* just as well as in the order in which we have placed them above. The *Coleoptera, Hymenoptera,* and *Diptera* are the three largest Orders, of each of which we have above three thousand representatives in Britain; of *Lepidoptera* we have two thousand species ; but the three other Orders are much less numerous.

The *Coleoptera,* or Beetles, have hard horny wing-cases, beneath which the wings are folded like a fan, and are then doubled over, so as to fit still closer. Their larvæ have six legs, and their pupæ are inactive and mummy-like, the legs of the future beetle being enclosed in separate sheaths. In some cases the perfect insect is destitute of wings and elytra, as in the female of the common glow-worm ; and in many other beetles the wings are absent, the elytra being either movable, but of course useless for flight ; or soldered together at the suture, as the line is called where the elytra meet over the back of the abdomen, but the elytra of beetles very seldom overlap. We will proceed to enumerate a few of the commoner and more interesting beetles.

The *Cicindelidæ,* or Tiger Beetles, are handsome, bright-coloured beetles, with large heads and strong jaws, which run and fly actively in the sunshine. The common Green Tiger Beetle (*Cicindela Campestris*) is green, with white markings, and is very abundant in many places. It feeds on other insects, and its larva is also carnivorous, forming a burrow in the sand, something like an Ant-lion, which is the name given to the larvæ of a family of the Order *Neuroptera,* a few species of which are found on the Continent, but which has no representative in this country. The Green Tiger Beetle is about half an inch in length.

The *Carabidæ,* or Ground Beetles, are also carnivorous. Several species of the typical genus *Carabus* are common, and may be seen running on paths or by the side of walls, especially in the morning and evening. They are oval beetles, about an inch long, and are of dark colours, with purplish, greenish, or brassy shades. They have movable elytra, but no wings. Many smaller species of ground beetles may be noticed in similar situations, most of them being black or greenish, often with bronzy reflections. One section frequents marshes, and a

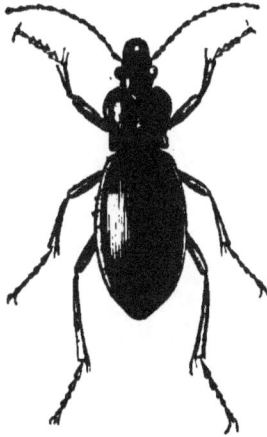

Ground Beetle (*Carabus Violaceus*), natural size.

few small species are found on the seashore at low-water mark, thus being among the very few insects which may be looked upon as marine.

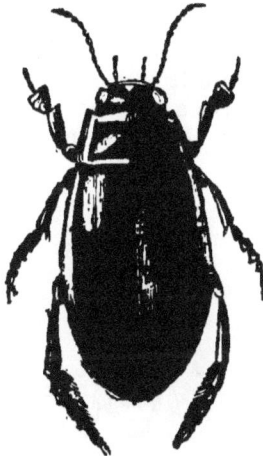

Dytiscus Marginalis (Male), natural size.

Although so few insects inhabit the sea, great numbers are found in fresh water, especially in their earlier stages ; but two or three families of beetles are more or less aquatic in all their stages. The species of *Dytiscus* are large, broad, flattish beetles, found in fresh water, and are of a brown or olive colour, with yellowish borders. They are very voracious, feeding on

Dytiscus Marginalis (Female), natural size.

smaller insects, and sometimes on small fish. They fly strongly, and in the evening often quit the water and fly to long distances. Many smaller species of *Dytiscidæ* inhabit our ponds and streams; but the most interesting of the smaller water beetles are the *Gyrinidæ*, or Whirligig Beetles, little black beetles with very long fore legs, which may often be seen rapidly circling about on the surface of the water.

The great group of *Staphylinidæ* may be known by their very

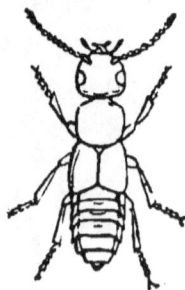

Devil's Coach-Horse (*Ocypus Olens*), natural size.

short elytra, which gives them a superficial resemblance to an earwig without the forceps. They may be found among all kinds of animal and vegetable refuse, though some are also met with on flowers. Some species inhabit the nests of ants, to which they often bear a superficial resemblance themselves. One of the largest and best known of the *Staphylinidæ* is the Devil's Coach-Horse, a large black insect with powerful jaws, and very rapacious. If alarmed, it lifts up its head and tail, and it is capable of inflicting a severe bite, which may be dangerous if the insect has lately been feeding on any putrid substance.

The species of *Necrophorus* (Burying Beetles) are black, with red markings on the elytra. They feed on carrion ; and if they meet with a mouse or bird, they dig a hole under it, and gradually pull and stamp it down into the ground, covering it up with earth. The female is buried with the carrion, in which she deposits her eggs, and then makes her way up to the surface of the ground again. *Hister* is a genus of small round beetles, of a shining black colour, with red spots. Their antennæ are clubbed at the tip, as is likewise the case in *Necrophorus*. Some smaller beetles allied to this, but of a more oval shape, belonging to the family *Dermestidæ*, are exceedingly destructive to hams, skins, and other dried animal products, *Dermestes Lardarius* and *Anthrenus Musæorum* having received their names from their food, or the localities in which they are found.

The largest water beetle found in England is *Hydrous Piceus*, which is half as long again as a *Dytiscus*. It is of a shining black colour, and is more convex and narrower than *Dytiscus*.

The *Scarabæidæ*, or Chafers, may be known by their short antennæ, the terminal parts of which are expanded into broad flat layers, which the insect can open or shut like a fan. All the species feed either on plants or on the dung of animals. They are large, broad, heavily formed insects, though some are very active on the wing. The Rose Chafer (*Cetonia Aurata*) is a bright green beetle, with whitish markings on the elytra. It measures rather more than half an inch in length, and is nearly as broad. It is often found nestling in roses, whence it derives its name, but is just as frequently found upon thistle, elder, and other flowers.

The Cockchafer (*Melolontha Vulgaris*) is a larger insect. Its larva feeds on the roots of plants, and is often very destructive to grass-fields, whereas the perfect insect is equally destructive to trees. The Cockchafer is brown, dusted with white in the male, and the under surface is banded with black and white. Several smaller species are similarly destructive to plants, one of which (*Phyllopertha Horticola*) is called the "Buckwheat Beetle" in Germany, where it swarms on the flowers of that plant. But it is equally common and very destructive in

England also. It is about half the size of the Cockchafer, and
has brown elytra, and a bronzy-green head and thorax. Shak-
speare's "shard-horn beetles, with their drowsy hum" belong to
the genus *Geotrupes.* They are black beetles, about the size and

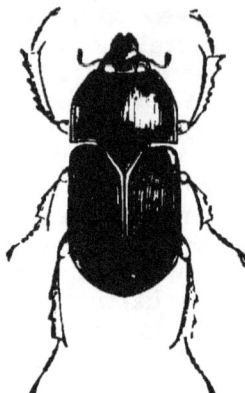

Geotrupes Stercorarius, natural size.

shape of a Cockchafer, and fly about in the evening. They are
black above and purple below, and feed on dung, as do likewise
the little beetles of the genus *Aphodius,* which resemble very
small Cockchafers in appearance, and are often to be seen flying
over dung, even on dusty roads.

The Stag Beetle is our largest British beetle, measuring nearly
two inches in length in large specimens, but it varies considerably

Click Beetle (*Elater Sanguineus*), natural size.

in size. It is remarkable for the enormous size of the jaws of
the male. The larva feeds on the wood of trees, and the
perfect insect on the sap.

The *Elateridæ* are long, narrow beetles, with hard wing-cases.

The hinder angles of the thorax are very acute, and many of the species have a habit of doubling themselves up when they fall on their backs, and jerking themselves on their legs with a click. They are therefore sometimes called "Click Beetles," and their larvæ are very long, slender, and tough, and are too well known to farmers and gardeners as wire-worms. The *Telephoridæ* resemble these beetles in their shape, being rather long insects : but they are of gayer colours, being reddish or ochreous instead, of black or bronzed, and their elytra are unusually soft for beetles. But their habits are predaceous, notwithstanding their apparently fragile structure. The Glow-worm (*Lampyris Noctiluca*), in which the male is a brown beetle, about half an inch long, and the female is wingless, is allied to these.

The *Tenebrionidæ* and allied families may be known from all the foregoing groups by having only four joints to the hind tarsi, and five on the front and middle legs. The preceding families have five joints to the tarsi of all the legs. They may also be known by their antennæ, which are moniliform, or composed of a number of bead-like joints. Several species are very familiar, such as the ugly, dull-black, wingless Cellar Beetles (*Blaps*); the rather narrow black beetle (*Tenebrio*

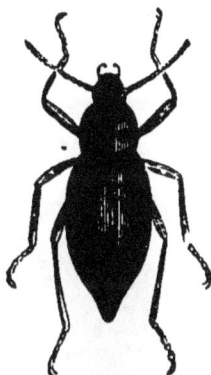

Cellar Beetle (*Blaps Mortisaga*), natural size.

Mollitor), the larva of which feeds on flour, etc., and is called the Meal-worm ; the large soft-bodied sluggish Oil Beetle (*Meloe*), which has no wings, and only rudimentary elytra, and which is generally found among grass ; and the beautiful green Blister Beetle (*Cantharis Vesicatoria*), which is found on ash-trees, but which is rare in this country.

The great group of *Curculionidæ*, or Weevils, may generally be recognised at once by their heads being produced into a kind of long tapering snout, near the extremity of which are placed the antennæ, which are often bent in the middle at a right angle. They are of all shapes, colours, and sizes, feed on different kinds of plants, and are sometimes very destructive. The Nut-Weevil is a familiar example; the larva is abundant

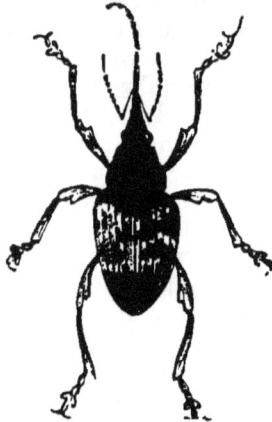

Nut-Weevil (*Balaninus Nucum*), magnified.

in nuts, and the perfect insect is a small brown beetle. In these, and several of the following families, the feet have only four joints to the tarsi.

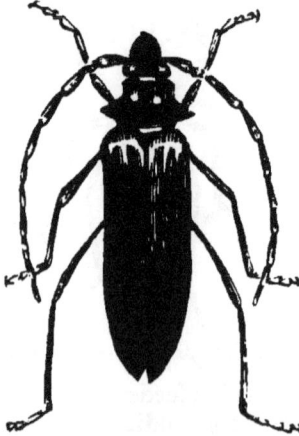

Musk Beetle (*Aromia Moschata*), natural size.

The Long-horn Beetles (*Cerambycidæ*) may be recognised by their very long antennæ. The commonest is the Musk Beetle, a green beetle about an inch long, which is found on the trunks of willow-trees, and emits a peculiar but agreeable odour. The larvæ of the *Cerambycidæ* feed for the most part on the wood of trees.

The *Chrysomelidæ*, or Golden-apple Beetles, are generally of a bright green or coppery colour, and are found, often gregari- ously, on the plants on which they feed. The Bloody-nose Beetle (*Timarcha Lævigata*), which is the largest British species, measures about half an inch in length, and is black, instead of green. It is of a roundish form, and very sluggish, and when touched it emits a reddish fluid; whence its popular name. The *Halticidæ* are smaller beetles, which possess the power of leaping. The destructive Turnip Fly (*Haltica Nemorum*), which is of a bronzy colour, with a yellowish stripe on each of the elytra, is the best known of this family.

The *Coccinellidæ*, or Lady Birds, have only three joints to the

Seven-spot Lady Bird (*Coccinella Septempunctata*), magnified.

tarsi. They are generally red or yellow, spotted with black ; and are very useful insects, as their larvæ feed on plant-lice. The *Trichopterygidæ*, the smallest known beetles, which are found among vegetable refuse, and are scarcely visible to the naked eye, are placed here by some entomologists.

The *Orthoptera*, or Straight-winged Insects, resemble the *Coleoptera* in the fore wings being much narrower and of a much thicker texture than the hind wings, which are the real organs of flight. The wing-cases are not horny, as in *Coleoptera*, but more resemble parchment. The metamorphoses of *Orthoptera* are imperfect, that is, the larva, pupa, and perfect insect re- semble each other, except that the larva is destitute of wings, which are rudimentary in the pupa. The pupa is active, and there is therefore no lengthened cessation of feeding or move- ment in the life of these insects.

The *Orthoptera* are rather poorly represented in cold countries; but the Order includes several common and well-known insects, as well as some of the largest and most conspicuous species that we possess.

Earwig (*Forficula Auricularia*), magnified.

The Earwigs (*Forficulidæ*) somewhat resemble *Staphylinidæ* among the *Coleoptera*, from which the pincer-like processes at

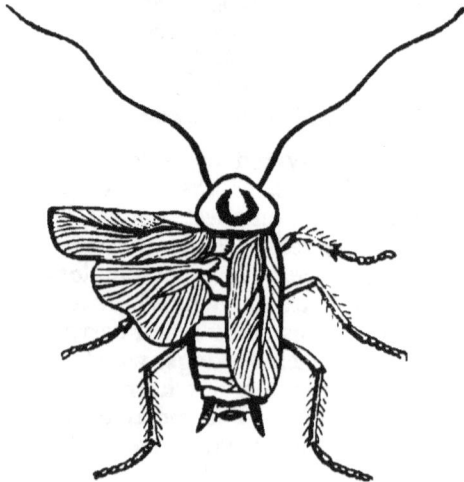

Cockroach (*Blatta Orientalis*), natural size.

the extremity of the body will at once distinguish them. They generally fly by night, hiding themselves in crevices during the day. Their wings are ample, but are perfectly concealed under the short wing-covers when not in use. They are very destructive to fruit and flowers—a hollow apple, or the flowers of the dahlia, sun-flower, etc., often harbouring a considerable number. The female is said to brood over her young like a hen.

The Earwig derives its name from its occasionally entering the ear—a fact which has been denied, but which is indisputable. The insect may be immediately dislodged from the ear by pouring oil into it.

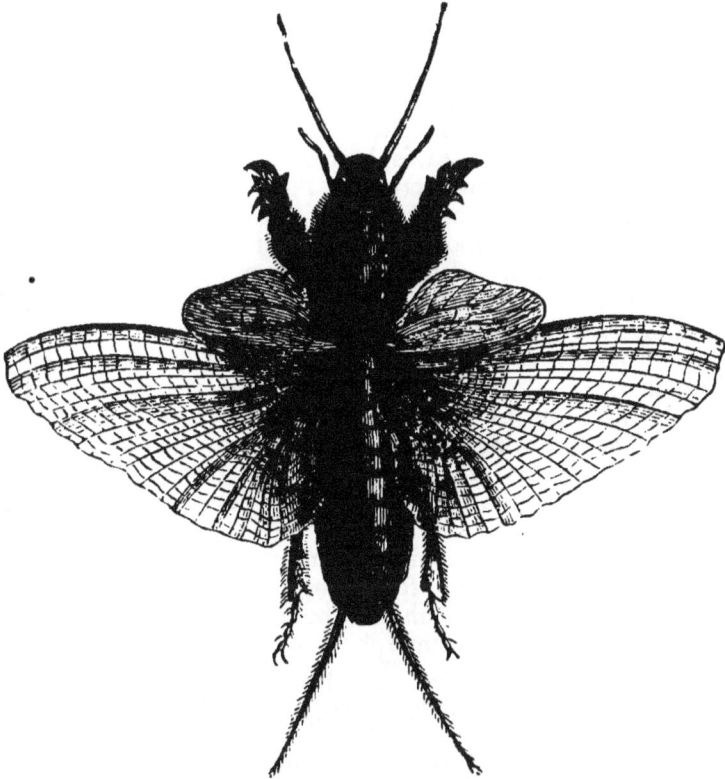

Mole Cricket (*Gryllotalpa Vulgaris*), natural size.

The Cockroaches are also nocturnal insects. The commonest species (*Blatta Orientalis*) is believed to be an importation from abroad, and is generally known as the Black Beetle. This description would apply well enough to *Blaps ;* but as *Blatta* is not a beetle, nor even black (being of a reddish brown), the popular name does not seem very appropriate. The eggs are laid in a single mass, being enclosed in a capsule ; as is also the case with locusts.

A less disagreeable insect is frequently associated with the Cockroach in town-houses—the Cricket (*Acheta Domestica*), which enlivens the winter evenings with its cheerful chirp. It is brown, but the Field Cricket (*Acheta Campestris*) is black. It is found on heaths and commons, though it is not nearly such a common insect as the House Cricket. The Mole Cricket (*Gryllotalpa Vulgaris*) is a much larger insect, which has the front legs formed nearly like the paws of a mole. It lives in burrows in the earth, and is rarely noticed above ground.

The Crickets may be distinguished from the Grasshoppers by their long antennæ. The Great Green Grasshopper (*Acrida Viridissima*), as it is called, is, however, really a cricket, though belonging to a different section to those mentioned in the last paragraph. It is of a bright green, the hind wings paler, and the female has an ovipositor about half as long as her body, which is rather short, although the wings expand about three inches. It is common in meadows, etc., in the south of England.

The Grasshoppers, which are about an inch long, belong to the same family as the Locusts, which they greatly resemble, except in size. The Migratory Locust (*Pachyteles Migratorius*) is about three inches long, and the wings expand about four inches. The fore wings are mottled with grey and light brown, and the hind wings are green. Stray specimens are not uncommonly met with in England, but it does not breed in this country.

The *Neuroptera* (Nerve-winged, or, more properly speaking, Net-winged Insects) have four wings of similar texture, not linked together by links on the borders. They are mostly carnivorous, and their metamorphosis is complete in some groups and incomplete in others. The principal insects included in this Order are the Dragon Flies, the Lace-winged Flies, the May Flies, and the Caddis Flies. The Dragon Flies are large, voracious insects, which live in the water during their earlier stages, where they undergo an imperfect metamorphosis, the pupa finally creeping out of the water, and giving birth to the perfect insect. The smaller species are generally found in the immediate neighbourhood of water ; but the larger and more active species are often found in woods and on heaths a long way from water. *Libellula Depressa* is a common species, measuring about three inches in expanse, and with a short, flat body, bluish in the male and yellow in the female. The giants of the section

B

however, belong to the genus *Æschna* and its allies. *Æ. Grandis* is a large brown species, measuring four inches in length, and even more across the wings, which are deeply tinged with yellow. The small species of *Agrion*, etc., with their slender blue, red, and brown bodies and delicate wings, often do not measure more than an inch and a half across the wings, and are generally found resting on the leaves or stems of water-plants.

Dragon Fly (*Libellula Depressa*), natural size.

The Lace-winged Flies (*Chrysopa*) are green flies, often found among bushes, with four transparent wings, something like those of a dragon fly ; but they are much smaller, measuring about three-quarters of an inch across the wings. Their bodies are slender, and comparatively short ; and notwithstanding their beauty, they emit a very disagreeable odour. Their larvæ feed on plant-lice (*Aphides*).

The May Flies (*Ephemera*) are found about running streams. They have long fore wings, about an inch and a quarter in expanse, and very short hind wings. Their tail ends in two or three long filaments, often quite as long as the body.

The genus *Perla*, including the Stone Flies, is intermediate in appearance between the May Flies and the Caddis Flies. The hind wings are nearly as long as the fore wings, the antennæ are long, the body rather broad and flattened, and the abdomen terminates in two long filaments. The larva is found under stones in rivers.

The Caddis Flies are sometimes formed into a separate Order, called *Trichoptera*, or Hairy-winged Insects. They are of different sizes, up to about two inches across the wings. *Phryganea Grandis* is the largest and one of the commonest species. They much resemble brown moths with rather narrow wings, which are clothed with hairs instead of scales, and their mouth is very imperfectly developed. Their larvæ live in water, where they form cases for themselves composed of bits of stick, stone, small shells, or any other material suitable for the purpose, which they can lay their hands on. They assume the pupa state in these cases, sometimes forming a slight cocoon.

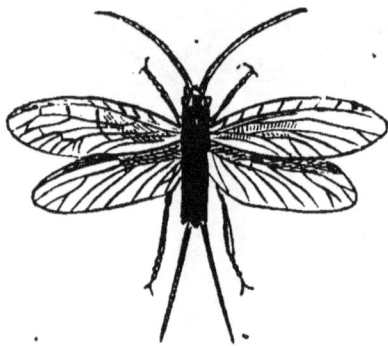

Stone Fly (*Perla Bicaudata*), natural size.

The Order *Hymenoptera* has four transparent wings, which are generally small in comparison with the size of the body. The fore and hind wings are linked together by a series of minute hooks on the borders. The female is always provided with a powerful ovipositor, which is frequently modified into a sting. The principal groups included in this Order are the Saw Flies, the Gall Flies, the Ichneumons, the Ruby-Tails, and the Ants, Wasps, and Bees.

The Saw Flies (*Tenthredinidæ*) derive their name from the boring apparatus of the female being modified into a pair of saws, which are used to cut a crevice in the bark or leaves of plants to receive the eggs. These eggs produce larvæ resembling those of *Lepidoptera*, but with from eighteen to twenty-two legs, whereas Lepidopterous larvæ never have more than sixteen. The best known species is the Gooseberry Fly (*Nematus Ribesii*), a yellow, four-winged fly, more or less spotted with black on the back, and measuring about half an inch in length. Its gregarious larvæ soon strip the gooseberry and currant bushes of their leaves, when they once get a footing in a garden.

The Gall Flies (*Cynipidæ*) are very small four-winged flies, which deposit their eggs in a somewhat similar manner under the cuticle of plants, and more especially on the oak. The puncture gives rise to an excrescence in which the larva lives and grows. These larvæ, however, are greatly infested with parasites, and you may sometimes rear several different species of small parasitic Hymenoptera from a gall, without the real owner being one of the party.

The *Ichneumonidæ* are a very large group of parasitic insects. Many of them puncture the bodies of caterpillars, and deposit an egg in each wound. The ichneumon larvæ, when hatched, devour the caterpillar alive, until it is full grown, or assumes the pupa state, when the larvæ quit the body of their victim,

Ichneumon Fly (*Pimpla Turionella*), Magnified.

frequently forming their cocoons around it. The Ichneumon Flies are often gaily coloured, with black and yellow markings. They are slender, elegantly formed insects, with long antennæ, and often a long ovipositor, which is sometimes formed of two or three filaments. There are several groups of *Hymenoptera* besides the Ichneumons proper which are parasites. Among these are the smallest of all, the *Proctotrupidæ*, or egg-parasites, to which we have already alluded.

The Ants, Bees, and Wasps frequently live in large communities, in which case they are the most intelligent of all insects.
The work of the nests is accomplished by undeveloped females,
called neuters, which form the bulk of the community.

In ants, the neuters are wingless, and the males and females
only acquire wings for their "marriage flight," after which the
males perish, and the few females which escape the pursuit of
their numerous enemies, either return to established nests, or
become the foundresses of new colonies.   Ants are far more numerous and annoying in hot climates than with us ; but the so-
called "White Ants," or Termites, which have very similar
habits, but are still more destructive, though happily not British,
belong to the Order *Neuroptera.* Several of our British ants
form nests in woods, fields, or gardens, and one little yellow ant
(*Myrmica Domestica*) is common in houses, where it is sometimes very annoying from its numbers.   Like nearly all our
noxious insects, it is an importation from abroad, and was almost
unknown fifty years ago.   It is believed to be a Brazilian species,

Red Ant (*Myrmica Rubra*), Male, Magnified.

which was first imported into the United States, and thence to
England.   Outdoor ants are very fond of the sweet substance,
called honey-dew, which exudes from the bodies of *Aphides*, or
plant-lice.   These they sometimes keep in their nests, sometimes tend on the plants where they feed, and sometimes even
superintend their breeding.   Many other insects are looked after
by ants in a similar manner, or are found in their nests ; and it is
no exaggeration to say that ants possess a much greater variety
of domestic animals than ourselves.   Concerning the metamorphoses of ants, I will only say here that they are most assiduous

in their attentions to their progeny, and that the so-called "ant-eggs" are not really eggs, but pupæ, which the ants expose to the proper amount of sun and air required for their development.

The solitary wasps and bees form nests in loose earth, or sometimes in decaying wood, differently constructed according to the species, and provisioned by the bees with honey, and by the wasps with caterpillars or other insects, which they sting in such a manner as to disable without killing them, so that a store of fresh provisions is always ready for the young larvæ when they hatch.

In the Humble Bees (*Bombus*) we already find small communities, consisting of perhaps a hundred individuals living together. There are no neuters among them, but the females differ very much in size, some being twice as large as others.

The Social Wasps belong to the genus *Vespa*, the species of which are all yellow with black markings, and are very similar to each other. They make their nests of a material resembling paper, either in the ground, under the eaves of a house, or suspended to the branch of a tree. The largest species, the Hornet (*Vespa Crabro*), generally forms its nest in a hollow tree. It is about twice the size of the other wasps, but much less common in most parts of this country, and its nests are much less populous.

Although, unlike bees, there are always many females in wasps' nests, yet every colony is founded by a single female which has survived the winter. Having constructed the beginning of a nest by herself, she continues her labours until she is joined by her progeny; and the whole colony works together to procure provisions and tend the young, until winter sets in. Then the wasps massacre the still immature larvæ and pupæ in the nest; and are themselves speedily killed by the increasing cold. A few females only survive the winter in a torpid state, to form fresh colonies next year. All the wasps which we see flying about in early spring are therefore females, each of which will soon found a new colony; and if we wish to diminish their numbers in summer, we can do so most effectually by destroying these wasps in spring.

The common Honey Bee or Hive Bee (*Apis Mellifica*) is scarcely to be considered wild, and has been introduced into every part of the world. A hive of bees contains one female, or queen bee; 200 or 300 males, or drones; and a large number of neuters, or workers, whose office it is to tend the larvæ and pupæ, construct the combs, and provide food for the com-

munity. The queen cannot bear a rival ; and whenever a queen
bee emerges from the pupa, a mortal combat ensues, the sur·
vivor becoming queen of the hive. But in spring, when the
hive becomes overcrowded, it generally happens that the old
queen and several of her successors rush out of the hive in a
huff, attended by a numerous escort ; and these become the
founders of a new hive. This is called " swarming."

The *Lepidoptera*, or Scale-winged Insects, include the Butter·
flies and Moths. They have four wings, clothed with a fine
dust which rubs off on the fingers, and which we find, under
the microscope, to be composed of elegantly formed scales.
They are classified primarily by differences in the structure of
their legs, wings, and antennæ. Butterflies fly by day, and have
ample, gaily coloured wings, and a more or less abrupt knob at
the end of their antennæ. Many butterflies have the front legs
more or less aborted, and useless for walking. As examples, we
may mention the Meadow Brown (*Epinephile Janira*), a brown
butterfly common in fields in summer, which has an eye-like
spot near the tip of the fore wings, surrounded with fulvous in
the female ; the Small Tortoiseshell (*Vanessa Urticæ*), a reddish
butterfly, with black spots on the fore wings, and a black border

Small Tortoiseshell (*Vanessa Urticæ*), natural size.

spotted with blue round all the wings ; and the Fritillaries,
which are fulvous butterflies spotted with black, and generally
with silvery spots on the under side of the hind wings. All
these are rather large butterflies, measuring an inch and a half
or more across the wings ; some of the larger Fritillaries ex·
pand nearly three inches. Our most delicately formed butterflies,
the Small Blue and Copper Butterflies, which we see flitting about
over flowers in waste places, only measure about an inch across

the wings.   They are particularly abundant in chalky localities;
and, notwithstanding their small size, are very pugnacious, often

Small Copper (*Lycæna Phlæas*), natural size.

driving other insects away when they approach the flower on
which they are resting.

The Brimstone Butterfly (*Gonepteryx Rhamni*), which is of a
sulphur-yellow colour, as its name implies, is very common
in woods in England, though almost unknown in Scotland or
Ireland ; it appears very early in the spring, and is to be found

Large White Butterfly (*Pieris Brassicæ*), natural size.

through a great part of the year.   It has an angular projection
on each wing, and thus differs from its allies, the White
Cabbage Butterflies (*Pieris*), three species of which are common
in our gardens, where their caterpillars feed on cabbage, etc.

The Skippers are small butterflies, about an inch in expanse.
They have rather thick bodies, large heads, and a rapid but
somewhat irregular flight.   Most of the species are brown with
fulvous markings, but the Grizzled Skipper (*Hesperia Malvæ*)
is black, chequered with white.   They frequent woods and
meadows,

We have only sixty-five different species of butterflies in England, but about two thousand moths. Out of this large number,
only a few can here be mentioned.

Large Skipper (*Pamphila Sylvanus*), natural size.

The *Sphingidæ* are large moths, with the antennæ thickened

Eyed Hawk-Moth (*Smerinthus Ocellatus*), natural size.

in the middle, long and rather pointed fore wings, and short
hind wings. Several species, as the Eyed Hawk-Moth, are very
beautiful moths. The largest of our British insects is the Death's
Head Hawk-Moth (*Acherontia Atropos*), which expands about
five inches. It has dark-brown fore wings, varied with paler,
and straw-coloured hind wings, with two black transverse stripes.
Its enormous green or brown caterpillar, with striped sides, and
a curved yellowish horn on the back near the extremity of the
body, is often found in potato-fields in autumn; but the Humming Bird Hawk-Moth (*Macroglossa Stellatarum*), which hovers

over flowers by day as well as at dusk, is more often noticed than any other species of this family.

The Six-spot Burnet Moth (*Zygæna Filipendulæ*) is common in meadows in summer. It has very thick antennæ, and measures an inch or more across the dark greenish fore wings, which are marked with six bright crimson spots arranged in pairs; the hind wings are short, and of a bright crimson. It flies heavily by day, and its larva constructs a tough yellowish boat-shaped cocoon, often found adhering to stalks of grass. The

Five-spot Burnet Moth (*Zygæna Trifolii*), natural size.

Five-spot Burnet Moth is very similar, but less abundant.

The Ghost Moth (*Hepialus Humuli*) flies in meadows in the evening. It measures about two inches across the wings, which are long, and comparatively narrow. The male is white above, and brown below; while the female has yellow fore wings with red markings, and dull reddish hind wings.

The Gold-tail and Brown-tail Moths (*Porthesia*) are common

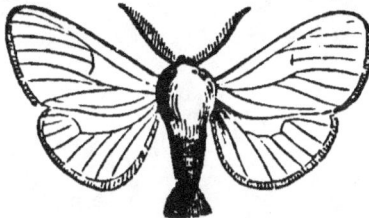

Brown-tail Moth (*Porthesia Chrysorrhœa*), natural size

in hedges in the evening. They measure about an inch and a quarter across the wings, which are snow-white, generally with a dusky spot near the hinder angle of the fore wings. They have a large tuft of down at the extremity of the body, which the female plucks off to cover her eggs.

The Tiger Moth (*Arctia Caja*) is a very handsome moth, often found in gardens. It measures over two inches across the fore wings, which are black, with irregular white stripes. The hind wings and abdomen are red, with black spots. Its caterpillar is sometimes called the Woolly Bear, and is covered with long red or black white-tipped hair. It rolls itself into a ball when disturbed, and feeds on a variety of low plants.

Yellow Underwing (*Triphæna Orbona*), natural size.

The dull-coloured moths, with stout and moderately long bodies, which fly in gardens in the evening, represent the night-moths (*Noctuidæ*) *par excellence*, and the caterpillars in some species do much harm by devouring the roots of plants. But they are not all dull-coloured, thus the Yellow Underwing (*Triphæna Orbona*) has brownish fore wings and yellow hind wings, with a black band near the hind margin; it expands about two inches.

Magpie Moth (*Abraxas Grossulariata*), natural size.

The Magpie Moth (*Abraxas Grossulariata*) may be taken as the representative of another extensive group of moths, the *Geometridæ*. It is often found in gardens, and has a slender yellow body, spotted with black, and broad white wings, expanding about an inch and a half; the fore wings are spotted with black and yellow, and the hind wings with black. Its caterpillar feeds on currant, and, like all of this section, has only ten legs instead of sixteen, the first two pairs of the abdominal legs, or prolegs, being wanting. This causes it to walk in a peculiar manner, forming an arch with its body at every movement.

The Small Magpie Moth (*Botys Urticata*), which is common among nettles, may be taken as a representative of the *Pyralidæ*. It expands rather more than an inch, and is white, with transverse rows of large blackish spots on all the wings; the thorax and base of the fore wings are yellow, the former with a few black dots.

The *Crambidæ*, or Grass Moths, have narrow, whitish, straw-coloured, or brown, white-streaked fore wings, and broad, brown hind wings, which they can fold into a very small space when at rest. They may also be known by their long palpi, two organs which project in front of the head in a kind of beak.

If we shake an oak-tree in summer, we shall probably dislodge a shower of small moths, about three-quarters of an inch in expanse, with broad green fore wings, and brown hind wings. This is *Tortrix Viridana*, and belongs to the *Tortricidæ*, a large group of small moths with broad fore wings, whose caterpillars feed in the rolled-up leaves of plants, or else in fruits, seeds, flower-heads, etc.

The *Tineidæ*, to which the Clothes Moths belong, are the most extensive group of the British *Lepidoptera*, of which they form nearly a third. They are all small moths, some very small, many measuring only a quarter or half an inch across the

Clothes Moth (*Tinea Tapetzella*), natural size.

wings. Their wings are generally long and narrow, and fringed with long hairs. Notwithstanding their small size, many species are adorned with brilliant metallic spots. One small family, the *Adelidæ*, includes green or brown species, with very long antennæ, which in some cases are at least an inch and a half in length, and double the expanse of the fore wings. Many of the larvæ of the *Tineidæ* feed inside leaves, where they form blotches or galleries.

The Plume Moths (*Pterophoridæ*) generally expand nearly an inch, and the wings are cleft into separate feathers; two, more or less united, on the fore wings, and three on the hind wings. In the *Alucitidæ* each wing is cleft into six distinct feathers; but of this family we have only one British representative. the

Macroglossa Stellatarum.

Saturnia Pavonia Minor.

Bembyx Mori.

[*Face p.* 28.

Twenty-plume Moth, a brownish insect, common in gardens, outhouses, etc., which measures rather more than half an inch across the wings.

The Order *Hemiptera* includes the Bugs and Frog-hoppers, and belongs to the Haustellate group of insects. Its members feed on the juices of animals or plants, which they imbibe through a powerful sucking apparatus. Their wings are more or less coriaceous or transparent ; and in the more typical species, from which the Order derives its name, the fore wings are

Plume Moth (*Pterophorus Lithodactylus*), natural size.

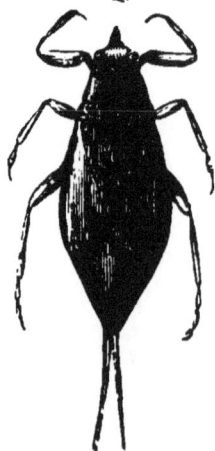

coriaceous at the base, and transparent, like the hind wings, at the extremity. Their metamorphosis is incomplete; and some species, like the common Bed Bug, are always apterous, while others exhibit both an apterous and a winged form. The antennæ are frequently short, being composed of only a small number of joints. The *Pentatomidæ* are very solidly formed, compact-looking insects, with five-jointed antennæ, and the scutellum, a plate which extends backwards behind the thorax,

which is generally small in other insects, extending in a triangular or nearly pentagonal form between the wings for two-thirds of the length of the abdomen. These bugs are found on plants ; but they feed on other insects, as well as on vegetable juices. There are many other Plant Bugs, some of which are long and slender insects, not at all like the heavy-looking *Pentatomidæ*. Most bugs exhale a special odour, different in different species, and not always unpleasant. The most disagreeable species next to the Bed Bug is the Wheel Bug (*Reduvius Personatus*), a black, rather narrow insect, about an inch long, which is common in outhouses. It feeds on other insects, and the larva and pupa conceal themselves from their prey by covering themselves thickly with dust. Among the

Water Scorpion (*Nepa Cinerea*), natural size.

Water Bugs, the Water Scorpion (*Nepa Cinerea*) may be easily recognised by its peculiar shape. It is black, with a red abdomen. The different species of *Notonecta* have grey fore wings and transparent hind wings. They have long hind legs, with which they row themselves about on their backs.

The second section of *Hemiptera*, called *Homoptera*, includes the *Cicadidæ* (of which we have only one rare species in England), the Frog-hoppers, Plant-lice, Bark-lice, etc. The Frog-hopper (*Aphrophora Spumaria*) is about half an inch long, and has greyish fore wings and transparent hind wings, and a broad head. . It flies as well as leaps ; and its larva is found in a mass of froth attached to the stems of grass, etc. ; this substance being known in country places as " cuckoo-spit." The *Aphides*, or

Cercopis Sanguinolenta (Magnified).

Plant-lice, are small insects of a brown or green colour, some of which are wingless, and others provided with large wings. They are very destructive to plants, exhausting their vitality by sucking the sap ; and they exude a sweet substance called " honey-dew," of which ants are very fond. They are popularly called " Smother Flies " ; and different broods of the same species exhibit very different forms, some colonies being wingless, and living at the roots of plants, while their progeny are winged, and infest trees or plants in the open air. Many broods consist entirely of females capable of reproducing their kind ; while other broods of the same species consist both of males and females.

The *Coccidæ*, or Bark-lice, also infest plants, but their habits are very different, the body of the wingless female forming a scale-like covering for her eggs after she has finished laying. In some cases, as in the Apple Blight, the female is covered with a cottony down. The dye called Cochineal is yielded by an exotic species.

The last Order of insects is that of the *Diptera*, or Two-winged Flies, in which the second pair of wings is rudimentary, consisting of a pair of organs resembling drum-sticks, called balancers or poisers. Only a few representatives of this extensive Order can be noticed here. They are suctorial insects in their perfect state, and feed on a great variety of animal and vegetable substances. Their metamorphoses are complete.

The Gnats (*Culicidæ* and *Chironomidæ*) are small, delicately formed flies, often with feathery antennæ in the males; and most of their larvæ live in water. In the Crane Flies or Daddy Long-legs (*Tipulidæ*), the larvæ live on the roots of grass and other

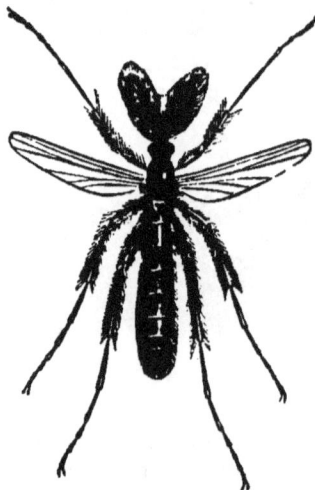

Chironomus Plumosus (Magnified).

plants, and are often **very** destructive. The legs of these flies are very long and slender, and break off with the slightest touch. Their wings expand from one to two inches, and are generally transparent, but sometimes variegated with brown. The female may be known by her short, horny ovipositor.

Syrphus Grossulariæ (Magnified).

One family of small flies (the *Cecidomyiidæ*) produces galls in plants, like the *Cynipidæ*.

The *Tabanidæ* are large flies which attack cattle, to which they cause great annoyance. Some species will also settle on man ; and it is quite common to feel a sharp sting on your hand in a wood, and see a pretty green fly, rather larger than a house-fly, with golden eyes and variegated wings, standing on your hand in a pool of blood.

Among the largest of the *Diptera* are the *Asilidæ ;* hairy flies, with thick legs, which feed on other insects. *Asilus Crabroni-formis* is an inch long ; the wings are tinged with yellowish, and the body is black, with the hinder half of the abdomen yellow.

The *Syrphidæ* are pretty flies, often chequered with black and yellow, which may be known at once by their peculiar hovering flight, darting away like an arrow when disturbed, and hovering again at the end of their flight.

The *Muscidæ* are a very large group, the larvæ of which generally feed in decaying vegetable or animal matter. ·The common house fly and the Blue Bottle may serve as examples.

The *Œstridæ*, or Bot Flies, are still more injurious to cattle than the *Tabanidæ*, as their larvæ live under the skin or in the stomach of our domestic animals. They are large, stout flies ; but many of the *Hippoboscidæ* and *Nycteribidæ*, which include the Forest Flies, Bird Flies, Sheep-ticks, and Bat-ticks, are apterous ; and their long, hairy, spider-like legs give them a superficial resemblance to some of the eight-legged parasitic *Arachnida*, which are not true insects, and to which the name Tick more properly belongs. The Fleas (*Pulicidæ*) are now considered by most entomologists to be a wingless and aberrant family of *Diptera*.

BOOKS ON ENTOMOLOGY.

The following books may be recommended as treating of Insects in general, without confining themselves to any special Order :—

Staveley's British Insects. (Coloured Plates).

Dallas' Elements of Entomology. (Woodcuts).

Westwood's Introduction to the Modern Classifica-tion of Insects. 2 vols., woodcuts (out of print, and only to be met with occasionally).

Pashord's Guide to the Study of Insects. (Wood-cuts). Treats chiefly of North American insects.

Kirby and Spence's Introduction to Entomology. (Treats of habits, etc.).

Ormerod's Injurious Insects. (Woodcuts).

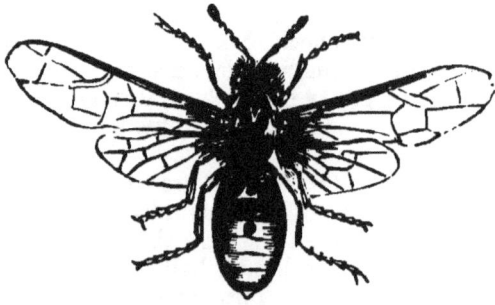

Cimbex Lutea.

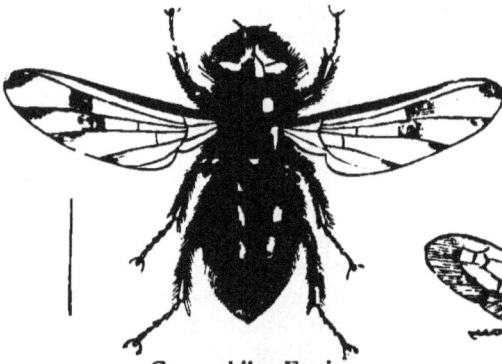

Gastrophilus Equi.

Bombus Pratorum.

Tingis Pyri.

Cimex Lectularius.

[*Face p.* 32.

BRITISH BEETLES.

C

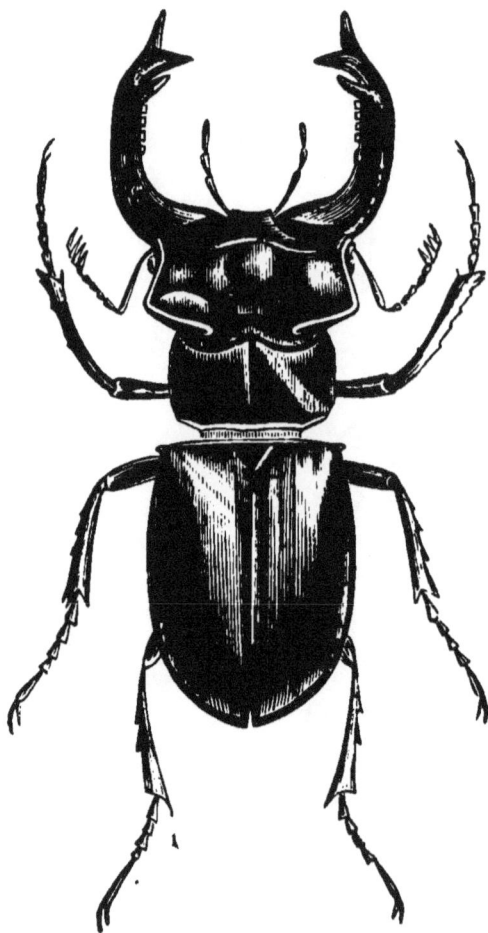

LUCANUS CERVUS (MALE), NAT. SIZE.
STAG-BEETLE.

BRITISH BEETLES.

MOST Entomologists commence the study of insects as boys, and begin by collecting everything they meet with; but they speedily find the subject too large to attack as a whole, and either abandon it entirely, or finally restrict their attention to one group of insects, their choice being commonly guided by the bent of some older friend, who has already formed a preference for one particular Order.

As the insects most easily collected and preserved are *Lepidoptera*, or Butterflies and Moths, and *Coleoptera*, or Beetles, it is only natural that these should receive the lion's share of the attention of Entomologists, though this is less exclusively the case than it was some years back. In the present little book it is our intention to give a few hints for the collection and preservation of beetles, followed by a brief outline of their classification, and a notice of some few of the more interesting British species.

On account of the greater number of *Coleoptera* often collected, their tenacity of life, and other reasons, it is found impracticable to pin them into collecting boxes in the field, as is generally done with *Lepidoptera*. Those with dark or metallic colours may be collected in weak spirits of wine, but this must be avoided in the case of red or hairy beetles. Or small beetles may be brought home alive, in a glass bottle with a little blotting-paper at the bottom, and a wide quill closed by a plug thrust through the cork, through which fresh captures may be dropped without opening the bottle. But larger or predaceous beetles must be put into another bottle, charged with some substance which will kill or stupify them at once. For this purpose some collectors use the young shoots or leaves of the laurel, which must be gathered when perfectly dry, and chopped fine ; others employ blotting-paper soaked in benzole ; or bisulphide of carbon may be used in the same way, though its extremely disagreeable smell is an objection. Many collectors use a bottle charged with a mixture of cyanide of potassium and plaster of Paris ; but the other methods mentioned above are just as useful in practice, and the use of cyanide (a most deadly poison) is better avoided.

The readiest way of killing beetles brought home alive (except soft-bodied, delicately-coloured, or downy beetles) is, however, to plunge them into boiling water, or to throw boiling water over them, as recommended by the Rev. Mr. Fowler, in a recent number of the *Entomologist*.   The same article contains other useful hints, not the least useful being that the sweeping and water-nets should be furnished with small rings round their upper edge, so that they can be slipped on and off the ring of the net at pleasure ; thus rendering it quite unnecessary to carry a large amount of apparatus.

The implements necessary for collecting beetles are, as just mentioned, a sweeping and a water-net, which can be combined, and a white umbrella. The ring should be made of strong galvanized wire, and of any convenient size, say about 9 inches in diameter ; and the nets may be composed of any strong white substance, not liable to tear by catching in brambles, and sufficiently close in texture to prevent small beetles escaping through it.   The water-net is always made of stronger material than the sweeping-net, but must not be " waterproof," as it is important that the water should run off readily.   The sweeping-net is used for brushing grass, trees, and bushes, from which many beetles will be swept ; and the water-net is used for fishing for water-beetles, either by dipping them up when you see them, or by drawing the net along the bottom of pools or streams in search of them.   The nets may be about twice as long as the diameter of the ring, and all the corners should be rounded off.   The stick should be strong, and of any convenient length.   The umbrella is used to beat or shake trees, bushes, or long grass over, when the beetles which fall can be picked up, and those which are too small may be lifted with a wetted finger.

Fungi, tufts of moss, or pieces of rotten wood, may also be examined over the umbrella in the field, or may be put in a bag or a botanist's vasculum to be examined at home.   Many beetles will be found running on field paths, sometimes in the heat of the sun, and sometimes in the evening ; others, as already im-

Elaphrus Cupreus.

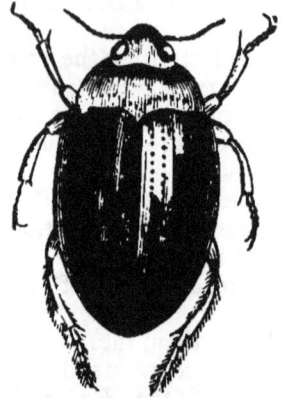

Haliplus Fulvus.

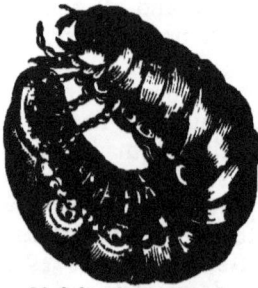

Melolontha Vulgaris
(larva).

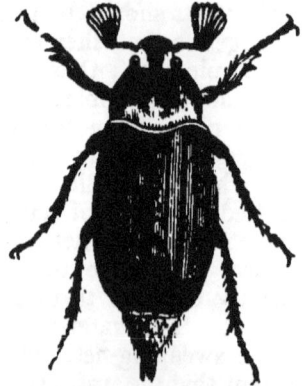

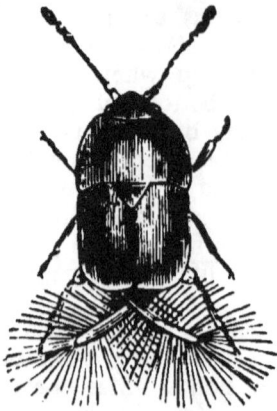

Melolontha Vulgaris.

Trichopteryx Atomaria.

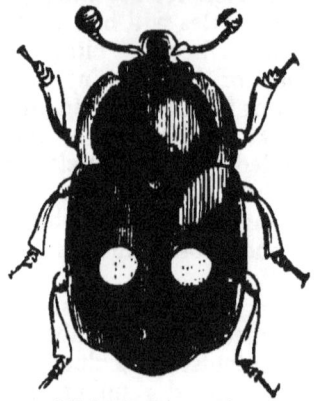

Nitidula Bipustulata.

[*Face p.* 38.

plied, frequent grass, flowers, and trees ; and others, again, feed on dung and other decaying substances. Many beetles fly little, and most of those which do have a heavy flight, and are very easy to capture.

Having collected your beetles, you have next to set them, and it is always better to do so as soon as possible after they are killed ; but if they have become too stiff, or if you are unable to set them immediately, they should be dropped into a jar half filled with chopped laurel, which must not be allowed to get mouldy. Those in spirit will keep for some time, but it is always better to set everything as soon as possible.

Large beetles must be pinned through the right wing-case, and their legs and antennæ spread out and kept in as natural a position as possible by means of pins, which must not be thrust through any part of the insect, but merely used to fix the limbs in position. Common pins are too thick and stiff for entomological use ; but proper insect pins can be purchased of any dealer in objects of natural history. It is not customary to spread the wings of beetles. Smaller beetles may be gummed on cards, their limbs arranged in a natural position, and left to dry. The gum used is prepared of gum tragacanth, to which a little gum arabic and acetic acid has been added. It is better to rule cards into sections of equal shape and size, and to mount a beetle on each ; they can then be cut up afterwards. The cards used in England are cut into an oblong form, but those employed on the Continent are long and pointed, the beetle being often mounted at the very tip. The former method has the neatest appearance in the cabinet. The boxes used to keep insects in are generally made double, like a backgammon board, and are lined top and bottom with cork. A little camphor or solid naphthaline must be kept in each box, to drive away mites. The collector will find it useful to keep some record of his captures, and the shortest way is to write a reference number beneath each card, or to stick a small ticket on the pin bearing a date corresponding with the entries in his journal, thus 1885, 7, 10 (July 10, 1885). This arrangement secures the utmost brevity, with perfect facility of reference ; for next year you will begin again in the same way, and thus avoid a long row of figures.

But now, having got together the nucleus of a collection of British beetles, how are you to begin to classify them? We have about 3,000 different kinds of beetles in this country, and at first sight it would seem to be a hopeless task to set about finding the name of any particular beetle ; yet, as everything known about beetles is registered according to the name of each,

we shall never be able to make any practical advance in our
knowledge of the subject, unless we are able, not indeed to name
every specimen offhand (for this is scarcely within the power of
the most experienced Coleopterist), but at least to assign it to
its place in the system with approximate accuracy.

First of all, how do we ascertain that our insect is really a
beetle ? Beetles belong to the Order *Coleoptera,* or case-winged
insects. They have four wings, like most other insects ; but the
two first are modified into stiff wing-covers called elytra, which
protect the delicate transparent under-wings, when these are
not in use, and serve rather as poisers than as locomotive organs
during flight. The elytra are generally hard and horny, though
sometimes of a leathery consistency, but always much stouter in
texture than the lower wings. They almost always meet down
the middle of the back in a straight suture. In some families,
especially in the *Staphylinidæ* and allied families, which are often
called *Brachelytra* on that account, they are very short, leaving
the greater part of the abdomen exposed. In some wingless
beetles the elytra are present and movable, and in others they
are soldered together ; while in a few instances, as in the female
of the common glow-worm, both wings and elytra are absent.

The character of the wings at once separates the *Coleoptera*
from all orders of insects except the *Orthoptera* and the *Hemiptera ;* but the last have a strong proboscis for imbibing their food,
whereas the *Coleoptera* are provided with mandibles for biting.
There remains the *Orthoptera ;* but in these insects, the tegmina,
as their wing-cases are called, differ much less from the lower
wings ; they are generally more or less veined, and often overlap
each other, in all which characters they differ from *Coleoptera ;*
besides, the *Orthoptera* have imperfect metamorphoses—that is,
the stages of larva, pupa, and perfect insect are not sharply
defined ; but in *Coleoptera* the metamorphosis is complete, and
a beetle passes through the four stages of egg, larva, pupa,
and imago, or perfect beetle.

The larvæ of beetles are white maggots, with a hard head, and
six legs, and the pupa is inactive ; but the cases which enclose
the various parts of the perfect insect are much more clearly
visible than in the pupæ of *Lepidoptera.*

Before proceeding to give a short sketch of the principal
sections into which the great Order *Coleoptera* has been divided
by Entomologists, it will be necessary to notice a few more points
in the structure of beetles. Their bodies are divided, like those
of other insects, into three principal portions; viz., head, thorax,
and abdomen. The most important parts of the head are the

eyes, the antennæ, and the mouth. There are two large compound eyes, one on each side of the head ; and two additional simple eyes, or ocelli, are occasionally present, placed on the top of the head. The antennæ, or feelers, which are long, jointed organs, are very important in classification, as they differ very much in structure in different families. They may be filiform, or thread-like ; moniliform, or bead-like ; pectinate, or feathery ; clavate, or knobbed at the tip ; lamellate, or furnished with a series of broad layers at the end opening out like a fan, etc. The mouth is composed of a variety of organs, which cannot here be described in detail; but we may mention the mandibles, or upper jaws, the maxillæ, or lower jaws, the labrum and labium, or upper and lower lip respectively, and the mentum, or chin. To the maxillæ and the labium are attached pairs of small feelers, called the maxillary and labial palpi respectively, and composed of only a few joints.

The thorax is composed of three segments soldered together, called the prothorax, mesothorax, and metathorax respectively. To the lower side of each is attached a pair of legs ; the elytra are attached to the mesothorax, and the wings to the metathorax. The legs are composed of the following principal parts : the coxæ, or hips ; the trochanters, or joints below the coxæ ; the femora, or thighs ; the tibiæ, or shanks ; and the tarsi, or feet. The tarsi consist of from three to five joints ; five is the normal number, and the principal exceptions will be specified. The abdomen requires little special notice here. In most beetles it is flattish above and convex below, and covered by the wings and elytra when these are closed.

The *Coleoptera* are divided into a great number of families and sub-families, which are classed together into larger groups, the first of which, the *Adephaga*, contains the carnivorous ground-beetles, and the water-beetles of the section *Hydrodephaga*. The ground-beetles, or *Geodephaga*, have long slender filiform antennæ, and powerful jaws. They are divided into two families, the *Cicindelidæ* and *Carabidæ*.

The *Cicindelidæ*, or Tiger Beetles, are very active predaceous insects, with large heads and eyes, and long slender

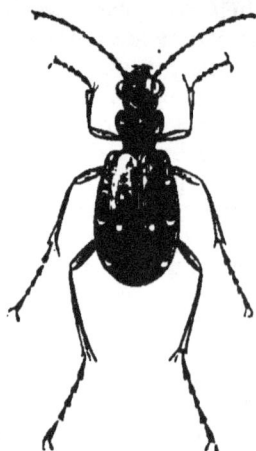

Tiger Beetle (*Cicindela Campestris*). (Mag.)

legs and antennæ. The beautiful green Tiger Beetle (*Cicindela Campestris*) is very common in sandy places, flying or running with great agility. Its larva forms a burrow in loose earth, very much like that of an Ant-Lion.

The species of *Carabus* are the largest of the *Carabidæ*, several measuring an inch or more in length. They are long, oval insects, and are black, often with . violet borders, or are more or less metallic in their colouring. The metallic species vary from dull bronze to brilliant green. Their wings are rudimentary, and they come out at dusk, and prowl about in search of prey during the night, though they may some-times be found running beneath walls, or along paths dur-ing the daytime, especially in spring. Many smaller species of *Carabidæ*, belonging to the genera *Pterostichus, Harpalus, Amara,* etc., are very common on paths in cornfields, and in similar localities. Many are black, bronzed, or green, while others are more brightly coloured. Although they are carnivorous, and probably de-stroy large numbers of injurious insects, yet it appears to be pretty well ascer-tained that they will sometimes attack corn; and one black species (*Zabrus Gibbus*), about half an inch long, is said to be very destructive to young wheat.

Harpalus Æneus.(Mag.)   *Harpalus Æneus*, a very common species, varies considerably in colour, but has always red legs and antennæ.

Many *Geodephaga*, and other beetles, will attempt to defend themselves by discharging a disagreeable acrid fluid when handled; but the Bombardier Beetles (*Brachinus*) have a more unusual mode of defence. They are small beetles, about a quarter of an inch in length, and are found under stones. The head and thorax are reddish, and the elytra are bluish or greenish. If they are alarmed, they discharge a slightly acid fluid, which immediately volatilizes into smoke, each discharge being accom-panied by a slight explosion.

Bombardier Beetle (*Brachi-nus Sclopeta*). (Mag.)

Many small species belonging to the genus *Bembidium*, though not exactly aquatic insects, frequent marshy places. They are generally black or bronzy, with yellowish spots and markings. The species of *Aëpus* are very small apterous yellowish beetles, found on the seashore at low-water mark; and as they are covered by the water for several hours every day, they may fairly be regarded as true

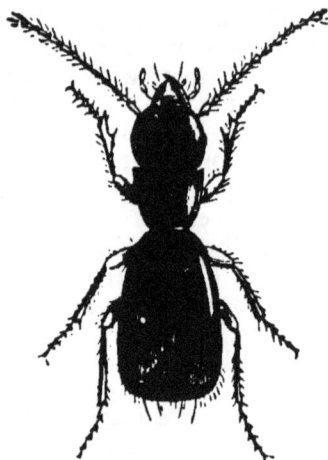

Aëpus Robinii. (Mag.)

marine insects. They are sometimes accompanied by *Aëpophilus Bonnairei*, a small insect of very similar appearance and habits, but belonging to the Order *Hemiptera*, or Bugs.

The *Hydradephaga*, or carnivorous Water-Beetles, include the two families *Dytiscidæ* and *Gyrinidæ*. The *Dytiscidæ* may be known from other water beetles by their long, slender antennæ, short palpi, and the structure of their legs. The front pair of legs is generally short, with the basal joints of the tarsi more or less dilated, at least in the males; in one group (the *Hydroporides*) the four front tarsi are only four-jointed. The largest and some of the commonest of the *Dytiscidæ* belong to the typical genus *Dytiscus*. They are olive green, or brown, with yellowish borders to the thorax and elytra. Water-beetles are generally very smooth and shining, but the elytra of *Dytiscus*, and several allied genera, are furrowed in the female.

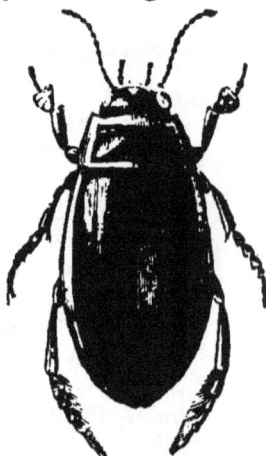

Dytiscus Marginalis. Nat. Size. (Male.)

The *Dytisci* are

Dytiscus Marginalis.
Nat. Size. (Female.)

very active and voracious insects, swimming about by day with the aid of their long hind legs, the tarsi of which are provided with a fringe of long hairs, and thus act as oars. These beetles will even attack and destroy small fish; and as they hybernate, they may be met with in the perfect state nearly all the year round. Their larvæ are also carnivorous, and are likewise aquatic; but the beetles leave the water in the evening, and sometimes fly to a great distance. The smaller species of *Dytiscidæ* are generally black or greenish, more or less marked with yellow; some of them are very pretty; and while some species prefer running water, others are more often found in stagnant pools.

The next family, the *Gyrinidæ*, or Whirligig Beetles, includes the most curious of all our water-beetles, as well as those which

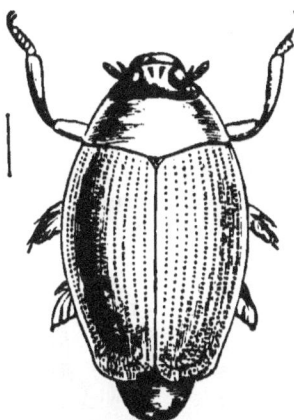

Whirligig Beetle (*Gyrinus Natator*). (Mag.)

most readily attract the attention of those who are not specially looking for them. They are oval, bluish-black beetles, about a quarter of an inch long, very smooth and shining, with very long front legs, and the two hinder pairs short and broad. They may be seen spinning in circles on the surface of the water throughout the summer. But the most remarkable peculiarity is in their eyes, which are completely divided in two, so that they may actually be said to have two eyes on the top of the head, and two on its lower surface, so that they can look upwards and downwards at the same time. There are very few other insects which possess so remarkable an apparatus.

The *Palpicornia* are a group of beetles which some authors place here, though others place them considerably further on. They include the two families *Hydrophilidæ* and *Sphæridiidæ*, and may be known by their short, clubbed antennæ, and their very long palpi, which are as long or longer than the antennæ. The

Hydrophilidæ are water-beetles, and one of them, *Hydrophilus Piceus*, is our largest water-beetle, measuring more than an inch and a half in length, though scarcely as broad as the *Dytisci;* *Hydrous Caraboides* is a smaller insect. Both these species are

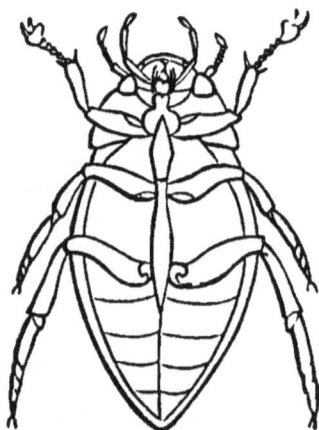

Hydrous Caraboides. (Mag.)

black and shining, and are found in stagnant water. They and their larvæ are carnivorous, feeding on small insects, etc., though the beetles are much less voracious than the *Dytiscidæ,* and feed to a greater or less extent on vege-table as well as animal matter. The females construct a silken silvery case to contain their eggs, provided with a tube at one end, which floats on the surface of the water, the egg-case itself being fixed to some plant. The smaller beetles of this family are mostly to be found in stagnant water among the roots of plants, though some few are not strictly aquatic. *Laccobius Minutus*, which we have figured, a common aquatic insect, is a little round black beetle, with pale yellow legs, mouth, and antennæ.

Laccobius Minutus. (Mag.)

The *Sphæridiidæ* are small round black beetles, sometimes spotted with red, which live in dung and under stones; a few species have yellowish elytra.

The *Brachelytra* are a very extensive group, differing very much in size and habits, but easily to be recognised by their very short elytra, which leave the greater part of the abdomen

exposed ; in fact, they much resemble an earwig in shape, except
the forceps.   This and the succeeding groups are divided into
so many families that our limits will only permit us to mention
some of the most important.   Many of the species of *Brachelytra* are black, often with a bluish or greenish reflection, but
others have the elytra, or their whole bodies, reddish or yellowish.

The *Aleocharidæ* are a very large family, including many
small species, which are found under stones and bark, in
marshy places among rushes, in fungi, under dead leaves, etc.
Many species are found in or near ants' nests, and we have
figured *Myrmedonia Collaris* in illustration, though it is not
a very common species.   It is a small insect, about one-sixth
of an inch in length, and is black, with the thorax and abdomen, except the tip, black.   What relation the beetles which

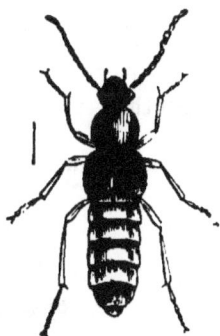

Myrmedonia Collaris.
(Mag.)

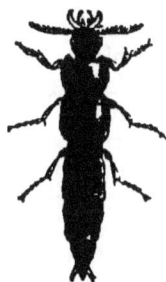

Velleius Dilatatus.
Nat. Size.

inhabit ants' nests bear to the ants, is not certainly known ; but
it is probable that they are employed in some manner as
domestic animals.   *Velleius Dilatatus*, which we have figured
as a representative of the *Quediidæ*, is a black insect, about
three-quarters of an inch in length, with reddish-brown antennæ.
It is nowhere common, but is generally met with in hornets'
nests, though found occasionally in the hollows of rotten trees
where no hornets are present ; its larva has some resemblance
to that of the hornet.

The largest of the *Brachelytra* is a large black beetle, popularly called the Devil's Coach Horse (*Ocypus Olens*).   It often

exceeds an inch in length, and feeds on dung or carrion. It runs about on roads, etc., by day, and, if touched, turns up its head and tail. It is armed with powerful jaws, and emits a very un-pleasant smell. This insect belongs to the family *Staphylinidæ,* and is common everywhere.

The *Clavicornia* are a rather extensive group, with clubbed antennæ, but their palpi are much shorter than in the *Palpicornia,* which some writers include in the present family.

The *Necrophoridæ,* or Burying Beetles, are black beetles, generally with transverse orange bands. They measure from half an inch to an inch in length, and the elytra are too short to reach to the extremity of the abdomen. They generally hunt in pairs, and if they find a dead mouse or bird, they carefully bury it by digging away the earth beneath, and pulling and stamping it down. After working for a day or two, with occasional intervals of rest, the male finally buries his mate with the car-case, in which she deposits her eggs, and then makes her way to the surface again.

Burying Beetle (*Necrophorus Rus-pator*). Nat. Size.

The *Silphidæ* are smaller and rounder insects, which likewise feed on carrion, and measure about half an inch in length. They are generally black, often with raised ridges on the elytra; but in *Silpha Thoracica* the thorax is reddish, and *S. Quadripunctata* has yellowish elytra, with two round black spots on each side. The largest species, *S. Littoralis,* more resembles a *Necrophorus* in shape; it is black, with the tip of the abdomen reddish. They may often be seen running on paths by day.

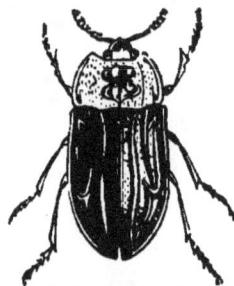

Silpha Thoracica. Nat. Size.

The *Histeridæ* are round black shining beetles, sometimes marked with red, which are found in dung, etc. Their form, and the very prominent club of the antennæ, render them rather conspicuous among our smaller beetles; they are about a quarter of an inch in length.

Many of the *Clavicornia* live in carrion, fungi, ants' nests, under bark, or in other situations where they are not only harm-less, but useful as scavengers; but the *Dermestidæ* are extremely injurious to hides, furs, etc.; the most destructive of all being

Dermestes Lardarius, the curious hairy larvæ of which will soon hollow out a ham, leaving nothing but the skin.   The beetle is

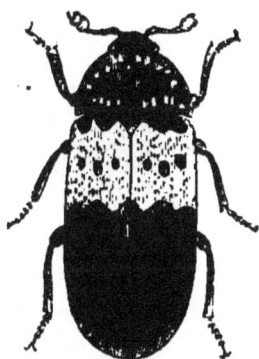

Dermestes Lardarius (Larva).   (Mag.)

Bacon Beetle
(*Dermestes Lardarius*).
(Mag.)

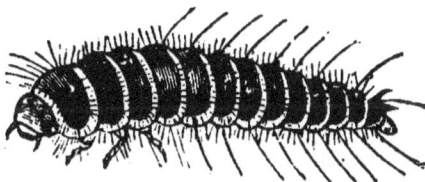

Dermestes Lardarius. (Pupa.) (Mag.)

black, with a broad brownish-grey band on the elytra, marked with three black spots on each side.

We have now arrived at the family *Lamellicornia,* which includes many of our largest and most conspicuous beetles. They derive their name from their antennæ terminating in a series of flat layers, which open or close at pleasure.

The largest of our British beetles is the Stag Beetle (*Lucanus Cervus*), (*vide* frontispiece), not an uncommon insect in the south of England, where it may be found in woods, resting on or near the roots of trees.   The larva feeds on wood, and the perfect insect feeds on the sap of trees, etc. ; it is said to saw off the ends of twigs with its jaws, by whirling itself round on the wing. The beetles, especially the males, differ considerably in size and in the development of the jaws ; and the female is able to give a sharper nip with her jaws than the male, although they are very small in comparison.

The Sacred Beetle of the Egyptians belongs to the family *Scarabæidæ.*   We have no species in England which has the peculiar ray-like teeth round the head ; but our nearest approach to it is *Copris Lunaris,* a black round beetle with a broad head, and a long horn in the middle in the male.

Typhæus Vulgaris, which belongs to the family *Geotrupidæ,* has also a short horn in the middle of the forehead ; the male is provided with three horns in front of the thorax, projecting forwards, and the tibiæ are strongly toothed.

The typical species of *Geotrupes* are roundish black beetles, often purplish beneath, which fly about heavily in the evening. They all feed on dung, like the two species last mentioned, which they much resemble, except that they have narrower heads, and no horns on the head or thorax.

The *Aphodiidæ* are also dung-beetles, but are more brightly coloured, many species being reddish or yellowish, or black varied with these colours. They are more oval, and much smaller than the other *Lamellicornia,* few of the species measuring as much as half an inch in length. The species of *Aphodius* are often seen flying about over roads. *Ægialia Arenaria,* which we have figured, is a blackish insect with undeveloped wings, which frequents sandy places, and is often met with on sandhills near the sea.

Geotrupes Stercorarius. Nat. Size.

Ægialia Arenaria. (Mag.)

The Cockchafer (*Melolontha Vulgaris*), typical of the family *Melolonthidæ,* is a large heavy-looking reddish-brown insect, more or less dusted with white ; the thorax is blackish, and the abdomen is black, with stripes of white hairs on the under surface ; the pointed tip of the abdomen projects beyond the elytra. The beetle measures rather more than an inch in length. It is one of our most destructive insects, for its white subterranean larva (which is generally found doubled up) feeds on the roots of grass, and the perfect insect feeds on the leaves of trees.

Another smaller but almost equally destructive insect is the Small Cockchafer (*Phyllopertha Horticola*), which belongs to the family *Rutelidæ.* It is bluish or greenish, with reddish-brown elytra, and measures less than half an inch in length. The larva feeds on the roots of plants, and the beetle feeds on flowers. It is called the Buck-wheat Beetle in Germany, where it hangs on the flowers of this plant almost in clusters.

The *Cetoniidæ,* of which we have only a few species in

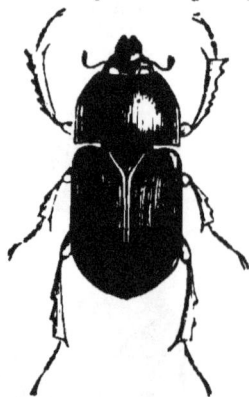

D

Britain, are also flower-loving beetles. The commonest species is the Rose Chafer (*Cetonia Aurata*), a beautiful green beetle, slightly marked with white on the elytra ; the under surface is a deep golden bronze. The larva feeds on rotten wood, and the beetle, which is very active in the sunshine, is found nestling in or flying around roses and other flowers. This insect is said to be used as a specific for hydrophobia in Russia ; and the statement has been repeated so long and so frequently, that it appears to deserve more serious attention than it has yet received.

The *Sternoxi* are rather long and narrow beetles, with hard elytra covering the whole abdomen, and with serrated or occasionally pectinated antennæ. The two principal families are the *Buprestidæ* and the *Elateridæ ;* in the latter the under surface of the prosternum has a projection behind, which fits into a hollow in the mesothorax.

The *Buprestidæ* are distinguished by the hinder angles of the thorax not being pointed, and by their not being able to leap. This group includes the splendid green beetles so common in the tropics, but is only represented by a few small green, blue, bronzy, or black species in England, the largest of which scarcely exceed a quarter of an inch in length. The larvæ of all the *Buprestidæ* feed on wood.

Click Beetle
(*Elater Sangui-
neus*). Nat. Size.

The *Elateridæ*, or Click Beetles, may readily be known by the hinder angles of the thorax being pointed, and by their power of jumping up, with a slight clicking noise, when laid on the back. Most of the species are black, or bronzed, or partly black and partly yellow ; *Elater Sanguineus*, which we have figured, is a bright scarlet insect, with a black head and thorax. The beetles are commonly met with on flowers, etc., in the daytime ; and their larvæ are too well known everywhere, as "wire-worms," being long and slender, with a very tough skin, and feeding on the roots of plants. The Fire Flies of South America are large species of *Elateridæ*, but I am not aware that any European species emits light.

The *Malacodermata* are not very unlike the *Elateridæ* in shape, but rather shorter. They have slender antennæ (rarely pectinated), and their elytra are generally very soft and flexible, and quite unlike the hard horny elytra of most other beetles ; several of the less typical families, however, have hard integuments.

The *Telephoridæ* are the most typical representatives of this group. The species of *Telephorus* are black, brown, or yellowish beetles, about half an inch long. They abound on the flowers of Umbelliferæ, etc., and are very rapacious, feeding on other insects. *T. Fuscus* is a common brown species, with the base and front of the head, the collar, and the edges of the abdomen reddish.

Telephorus Fuscus. Nat. Size.

The *Lampyridæ* are also carnivorous, but differ from the *Telephoridæ*, in the females being apterous. *Lampyris Noctiluca* (the Glow-worm) is common in many parts of England. The male is greyish-yellow, and about half an inch long; it flies by night, and is very slightly luminous. The female is completely apterous, and may easily be detected among the grass by its light.

Drilus Flavescens is another insect which some authors class

Drilus Flavescens. (Mag.)
(Male.)

Drilus Flavescens. (Mag.)
(Female.)

with the *Lampyridæ*, while others regard it as belonging to a separate family, *Drilidæ*. Both sexes much resemble the corresponding sexes of the Glow-worm in appearance, but are only half the size, and are not luminous. The larva feeds on snails, and forms its pupa in the shells.

Several of the smaller beetles with hard integuments which are classed with the *Malacodermata*, feed chiefly on wood. These belong to the genera *Ptinus*, *Anobium*, etc., and form the family *Ptinidæ*. Several species are found in houses, the best known being *Anobium Domesticum*, a small brown beetle about one-sixth of an inch in length, which is found in timber, furniture, etc., and produces a slight noise as a call to its mate. Its popular name is the "Death-Watch."

We have now to consider the *Heteromera*, a group of beetles easily recognisable by two very obvious characters. The four front tarsi are five-jointed as usual, but the hind tarsi are only four-jointed. The antennæ, too, consist of a series of nearly round joints, thus resembling a string of beads.

The family *Blaptidæ* only includes three British species, which are found in cellars, stables, etc. We have figured *Blaps Mortisaga*, which is the rarest; but the other species, *B. Mucronata* and *B. Similis*, very much resemble it. All three are black, wingless insects, with the elytra soldered together at the suture, and immovable. Other species of *Heteromera* are found in houses, such as *Tenebrio Mollitor* (the Meal Worm), which is a black beetle, about half an inch long, but winged, and rather long and narrow; it belongs to the family *Tenebrionidæ;* its larva feeds on flour, etc.

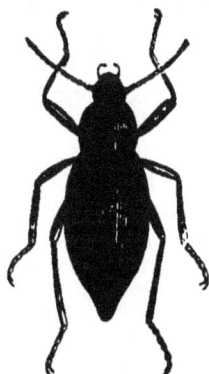

Cellar Beetle
(*Blaps Mortisaga*).
Nat. Size.

The *Pyrochroidæ* are remarkable for their bright red or scarlet colour, although many of the *Heteromera* are very dingy in appearance.

The *Meloidæ*, or Oil Beetles, are very strange-looking insects, perfectly incapable of flight, having no wings, very short, soft elytra, and very clumsy, bloated-looking bodies. They are common in grassy places in spring, and their larvæ are parasitic in the nests of bees. They are large, blue-black beetles.

The Blister Beetle (*Cantharis Vesicatoria*), belonging to the family *Cantharidæ*, is a most brilliant green beetle, rather more than half an inch in length. It is not common in England, though sometimes met with on ash and other trees. The beetles used in medicine are brought from South Europe, where they are abundant; and I have not seen any beetle which presents so brilliant an appearance as this, when the sun is shining on a tree on which several of them are feeding.

Blister Beetle
(*Cantharis Vesicatoria*). Nat. Size.

The next group of beetles is that of the *Rhynchophora*, or Weevils. It is one of the most extensive, and its members may be known immediately by the head being long and pointed in front, forming a kind of beak, on each side of which the antennæ are placed, which are angulated in the middle, and clubbed at the extremity. They have only four visible joints to all the tarsi (which is likewise the case in the *Longicornia* and *Eupoda*), and their integuments are much harder than those of most other beetles.

The *Bruchidæ* are injurious to peas and beans, and we have figured the Pea Weevil as an illustration of the group. The beetles are black, with white pubescence, and are about one-sixth of an inch long. They appear in spring, and lay their eggs when the pod is quite young ; and when the larva is hatched, it devours the peas. The rostrum in the *Bruchidæ* is very short, and the antennæ are not angulated ; we rarely find the characters of a group exhibited in perfection by the first or last families which are included in it. The peas and beans which are infested by these insects are extremely injurious to the animals which feed upon them.

Pea Weevil
(*Bruchus Pisi*).
(Mag.)

Apion Flavipes. (Mag.)

The *Apionidæ* are an extensive family of small weevils, many of which do not exceed one-tenth or one-twelfth of an inch in length. They are black, blue, green, or red, with broad elytra, but no wings, and a narrow head and thorax, the former produced into a long rostrum, and the antennæ inserted about the middle. They are found gregariously on various plants on which the larvæ feed. *Apion Flavipes*, which we have figured, is black, with reddish legs ; it is found on trefoil.

Rhynchites Bacchus, which belongs to the family *Rhinomaceridæ*, is a beautiful little purple beetle, with a golden lustre ; it is about one-sixth of an inch in length. It is met with in spring on apples and sloes, devouring the buds ; and later in the year it deposits an egg

*Rhynchites
Bacchus.*
(Mag.)

in the young fruit. The larva eats its way out in three or four weeks, and forms its pupa in the ground. It is, however, too scarce an insect in England to be very destructive. There are several other species of the genus, some of which injure fruit in the same way, and others feed on the leaves of various trees, sometimes rolling them together, and laying an egg in them, and sometimes injuring a young shoot till it withers, and then laying an egg in the pith.

Balaninas Nucum (the Nut Weevil), belonging to the family *Erirhinidæ*, is a black beetle about a quarter of an inch long, with dull red legs, and a long proboscis or rostrum, thick at the base, near which the antennæ are placed. It springs from the white larva that we so often find in nuts, especially filberts ; but

it will also attack acorns, and there are several allied species which will likewise feed on both fruits.

Cryptorhynchus Lapathi, typical of the family *Cryptorhynchidæ*, is a larger beetle, one-third of an inch in length, with a shorter and more gradually formed rostrum. It is a convex, dull black insect, with the sides of the thorax, the base, and nearly half the extremity of the elytra, and the femora, thickly scaled with white. Its black and white colouring makes it rather a conspicuous insect. The beetle lays its eggs in May on young willows, and the larva feeds on

*Cryptorhynchus
Lapathi.*
(Mag.)

the pith. The beetle appears in autumn, and lives through the winter.

The *Scolytidæ* are the most destructive of all the wood-feeding beetles, for their larvæ gnaw long galleries through and through the trunks of trees, and speedily destroy them. They have scarcely any rostrum, and their antennæ are short and clubbed. *Hylesinus Piniperda* is a black or brown insect, about one-sixth of an inch in length ; the antennæ and legs are reddish. The beetle and its larva may be found beneath the bark of different kinds of fir almost all the year round. *Scolytus Destructor*, which feeds on elm, is perhaps the best known of this family.

Hylesinus Piniperda.
(Mag.)

The *Longicornia*, or Long-horned Beetles, may be known at

once from any other beetles with four-jointed tarsi, by their usually large size and their very long antennæ. Their larvæ are all wood-feeding insects, but are not generally sufficiently numerous to cause much damage. The larger beetles may sometimes be found resting on the trunks of trees by day, especially near the roots; and some of the smaller species are very active on the wing.

One of the largest and commonest of our British Longicornia, belonging to the typical family *Cerambycidæ*, is the Musk Beetle, *Aromia Moschata*, which emits a strong but rather agreeable odour. It is a long green beetle, measuring an inch or more in length, with curving antennæ at least as long as the body, and the sides of the thorax with a sharp angular projection. It is sometimes very common in summer on the trunks of willows, in which its larva feeds. It is not very active, and may easily be seized with the fingers, as it is quite harmless.

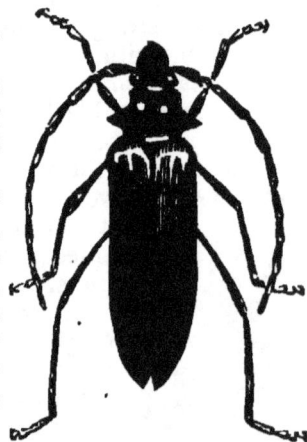

Musk Beetle (*Aromia Moschata*). Nat. Size.

We will now pass on to the large group of beetles called *Eupoda*. The majority are vegetarians, and hence the group is sometimes called *Phytophaga*, or Plant-feeders. They have four joints to the tarsi, like the *Rhynchophora* and *Longicornia*, and their antennæ are generally short and slender; their bodies are round or oval in shape.

The *Donaciidæ* have longer antennæ, legs, and bodies than most of the succeeding families, from which they likewise differ in their semi-aquatic habits. The species measure about one-third of an inch in length, and are generally bright green or bronzed, though some are purplish, or even black. Their larvæ feed under water on the roots of water-plants, not being very particular in their choice. The beetles are fond of basking on water-plants; the under surface of their bodies is clothed with a white down, which enables them to carry a bubble of air beneath the water when diving.

The *Crioceridæ* have oval elytra, shorter and broader than in the *Donaciidæ*, and the antennæ are rather short and thick. *Crioceris Asparagi*, which measures nearly a quarter of an inch in length, is a common garden insect. It is of a bluish-green, with a red thorax, and red edges to the elytra; each elytron is also marked with three more or less confluent yellowish spots or blotches.

The *Cryptocephalidæ* and *Chrysomelidæ* include more rotund beetles, with longer antennæ than *Crioceris*. Many are of a brilliant golden green, and are found gregariously resting on the various plants on which they feed. Some species are blue or blue-black, sometimes with a red border; and others, again, are black.

One of the most beautiful species is *Chrysomela Cerealis*, which is of a brilliant golden green, with a purplish lustre, and with three bands on the thorax and three on each elytra, besides the suture, of a deep blue, bordered with green. It is about one-third of an inch in length, and is found under stones in spring, and later in the year on grass and various low plants, but is not very common, though met with occasionally on the Welsh mountains. *C. Banksii*, a rather larger insect, of a bronzy-green colour, is very common among grass.

Although it is not our intention to include notices of foreign insects in the present series of elementary handbooks, yet an exception must be made in the case of *Leptinotarsa Decemlineata*, the dreaded Colorado Potato Beetle, which belongs to the *Chrysomelidæ*. It is about one-third of an inch in length, and the elytra are marked with alternate stripes of black and dull yellow; the thorax is also yellow, with a blackish V-shaped mark in the centre, and several dark spots on each side. But a very striking peculiarity which will at once identify the insect, is its wings, which are not colourless, as in most other beetles, but red. It has proved so destructive in North America that its introduction into England is strictly forbidden, and a farmer was lately fined five pounds for this offence.

The *Halticidæ* are a family of small beetles, which are too well known to the farmer, as the notorious Turnip Fly is one of their number. They are oval insects, often measuring less than one-twelfth of an inch in length, and have thickened femora, which enables them to leap almost like fleas. The species of *Phyllotreta* are black, generally with a yellow stripe on each elytron, and are all of similar habits, and about equally destructive to turnips, etc.

The *Cassididæ*, or Tortoise Beetles, are easily known by their flattened appearance, the thorax being developed in such a manner as to cover the head like a shield; the legs and antennæ are rather short and thick, and the beetles are very sluggish. They and their larvæ feed on low plants, which they strip to skeletons. The larvæ have a curious fork-like projection behind, which curves over their bodies. Upon this they pile their excrements, by which they are thus always overshadowed. Most of the species are green, but some are black ; and others are red,

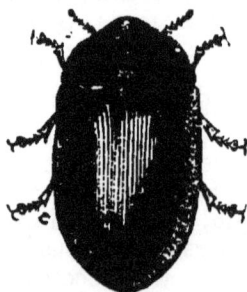

Tortoise Beetle
(*Cassida Oblonga*).
(Mag.)

with black spots. *C. Oblonga* is green above, with a golden band on each elytron ; the under surface is black. It is about a quarter of an inch in length, and is not an uncommon species.

The *Trimera*, which form the last group of beetles, may be known by having only three visible joints to the tarsi. The species inhabit fungi, ants' nests, etc. ; many feed on vegetable refuse ; while others, again, are carnivorous. Most of the species have clubbed antennæ, and many authors include part, at least, with the *Clavicornia*, in spite of the singular structure of their tarsi.

The *Coccinellidæ*, or Lady-birds, are small, smooth, round beetles, with red or yellow elytra, spotted with black. The commonest species is *Coccinella Septempunctata*, the Seven-spot Lady-bird, which has a black head and thorax, with scattered white marks, and red elytra, with three black spots on each, and one near the base on the suture. These insects and their larvæ are very useful to agriculturists, as they destroy the Aphides, otherwise known as Plant-lice, Smother-flies, or Blight, which do so much mischief to our cultivated trees and plants. Lady-birds are migratory when abundant, sometimes making a sudden and unexpected appearance in some special locality in enormous numbers.

The *Pselaphidæ* and *Truhopterygidæ* include the smallest

known beetles. They have clubbed antennæ, and the former have short elytra. They are found among moss, grass, under stones, in ants' nests, etc., and are of almost microscopic

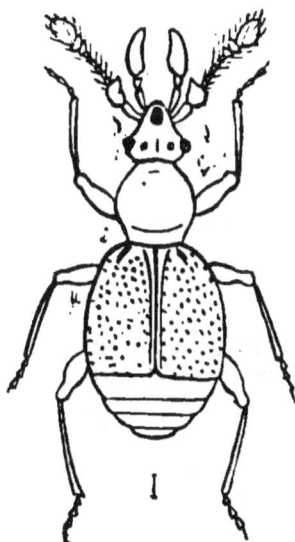

Bythinus Curtisii. (Mag.)

minuteness. The species figured (*Bythinus Curtisii*) is one of the *Pselaphidæ.* It is a brown beetle, with reddish brown legs and antennæ, and is not uncommon.

It is impossible within the few pages to which we are restricted to give more than a very brief outline of so extensive a group of insects as the British Beetles ; and those who wish to pursue the subject further will find more comprehensive works on the same subject by E. C. Rye and Herbert Cox. Calwer's "Käferbuch," a German work, with coloured plates, will also be found very useful for the figures, quite apart from the letterpress.

SKETCH OF THE PRINCIPAL GROUPS OF BRITISH BEETLES.

GEODEPHAGA.   Ground Beetles.

Carnivorous beetles, with five joints to the tarsi, and thread-like antennæ.

Cicindelidæ.   Tiger Beetles.

Green or brown beetles, with white markings; head and eyes large; legs long; very active by day; frequenting sandy places.

Carabidæ.   Ground Beetles proper.

Head and eyes smaller; less active; sometimes wingless; frequenting fields, marshes, etc.

HYDRADEPHAGA.   Carnivorous Water Beetles.

Dytiscidæ.   Water Beetles.

Antennæ slender, much longer than the palpi; hind legs formed for swimming; front legs short.

Gyrinidæ.   Whirligig Beetles.

Front legs long; four hind legs short.

PALPICORNIA.   Omnivorous Water Beetles.

Antennæ short, clubbed; palpi as long or longer than the antennæ.

BRACHELYTRA.   Rove Beetles.

Abdomen long, elytra very short; feed chiefly on decaying vegetable or animal matter.

CLAVICORNIA.

Antennæ clubbed; palpi much shorter than in the *Palpicornia*: feed on dung, carrion, etc.

LAMELLICORNIA.   Chafers.

Antennæ short, terminating in a club formed of a number of movable layers ; feed on plants or dung.

Lucanidæ.   Stag Beetles.

Mandibles of the male very large; larvæ feed on wood; beetles on sap.

Scarabæidæ.

Head very wide and flattened, with a horn in the middle in the male ; black ; feed on dung ; fly in the evening.

Geotrupidæ.

Head not of unusual size ; black ; feed on dung.

Aphodiidæ.

Small oval dung-beetles, of various colours ; fly by day.

Melolonthidæ.   Chafers.

Large beetles ; larvæ feed on grass ; beetles on the leaves of trees, round which they fly in the evening.

Cetoniidæ.   Rose-Chafers.

Brightly coloured or black beetles (sometimes downy) found feeding on roses, thistles, etc., by day.

STERNOXI.

Long and narrow beetles, with serrated or pectinated antennæ ; elytra hard ; feed on plants.

Buprestidæ.

Hinder angles of the thorax not pointed.

Elateridæ.   Click Beetles and Wire-Worms.

Hinder angles of the thorax pointed ; able to leap ; larvæ feed on roots of plants.

MALACODERMATA.

Long and narrow beetles, generally with slender antennæ and soft elytra; frequent flowers, but attack other insects ; female sometimes luminous and apterous (as in the glow-worm). (The *Ptinidæ*, however, are small oval, wood-feeding beetles with hard integuments.)

HETEROMERA.

Beetles with hard or soft elytra, sometimes wingless ; antennæ bead-like ; four front tarsi five-jointed ; hind tarsi four-jointed ; feed on vegetable substances.

RHYNCHOPHORA.   Weevils.

Snout very long, antennæ placed on each side, and elbowed; integuments hard ; all the tarsi four-jointed ; plant-feeders.

LONGICORNIA.   Long-horned Beetles.

Long and rather narrow beetles ; antennæ very long ; tarsi four-jointed ; feed on plants, their larvæ often burrowing in the wood of trees.

EUPODA.

Round or oval beetles ; antennæ of moderate length ; tarsi four-jointed ; feed on plants.

Donaciidæ.

Semi-aquatic ; feed on water-plants.

Chrysomelidæ.

Bright-coloured roundish beetles, found among grass, etc.

Halticidæ.   Turnip Beetles.

Small beetles with thickened femora, which leap like fleas ; very destructive to turnips, etc.

Cassididæ.   Tortoise Beetles.

Flattened beetles, with the thorax overlapping the head.

TRIMERA.

Small beetles, with only three visible joints to the tarsi.

Coccinellidæ.   Lady-Birds.

Small spotted beetles, which feed on plant-lice.

Trichopterygidæ.

The smallest known beetles ; antennæ clubbed ; found among vegetable refuse, in fungi, or in ants' nests.

ON THE PART PLAYED BY BEETLES IN THE ECONOMY OF NATURE.

A "Beetle" often conveys the idea of something peculiarly repulsive; but this perhaps arises chiefly from its improper association with the "Cockroaches," or "Black Beetles," of our kitchens, which, however, belong to the Order *Orthoptera*, and are therefore, strictly speaking, not "Beetles" at all.

An enormous number of beetles are now known; nearly 100,000 of all shapes, colours, and sizes, from a speck scarcely visible to the naked eye, to about six inches in length; and although beetles are not the largest insects in expanse of wing, some of them are perhaps the heaviest and bulkiest insects known.

They share with other insects the offices both of general scavengers and also of checks upon the too great luxuriance of vegetation. Many beetles feed on carrion, and especially small animals; but the work of clearing away such substances is chiefly accomplished by the larvæ of various two-winged flies (*Diptera*). Other beetles feed on dung, which they often attack the very moment it is dropped; and you cannot turn up a patch of dried dung without finding it swarming with beetles. The plant-feeding and wood-feeding beetles, according to their species, attack almost every portion of every plant; and any species feeding on a cultivated plant is likely to produce great destruction, if it becomes unusually abundant. Of the beetles which feed on other insects, the most important are the Lady-Birds, or *Coccinellidæ*, which destroy the *Aphides*. We do not meet with many parasitic insects among beetles, but the larvæ of the Oil Beetles, or *Meloidæ*, and those of the *Stylopidæ*, are parasitic on *Hymenoptera*. The *Stylopidæ*, which used to be placed in a separate Order (*Strepsiptera*), are hardly likely to fall under the notice of beginners. They are small black insects, with a single pair of very large wings in the male; and the larvæ are parasitic in the bodies of bees, which the apterous female never quits; and which the male only leaves on emerging from the pupa.

Beetles have probably not been made of so much use to man

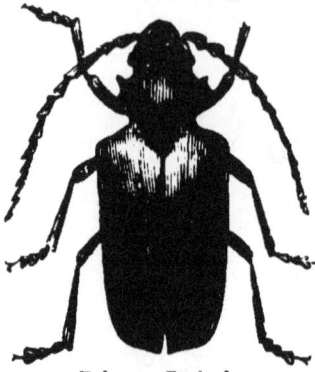

Prionus Coriarius.

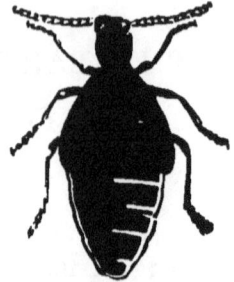

Meloe Proscarabæus.

Crioceris Asparagi.

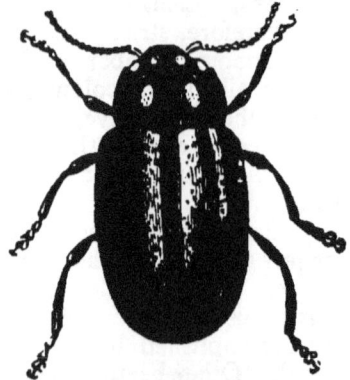

Chrysomela Cerealis

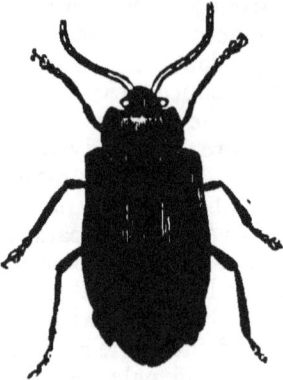

Galeruca Tanaceti.

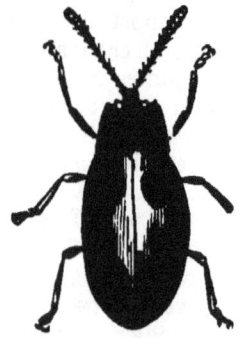

Endomychus Coccineus.

[*Face p.* 62.

as they might. A few species are eaten in different parts of the world. The Sacred Beetle is sometimes eaten by the Egyptians; and the larvæ of various large wood-feeding beetles are considered a great delicacy in the West Indies and elsewhere. The Blister Beetles and the Rose-Chafer are sometimes employed in medicine. The splendid green *Buprestidæ* of the Tropics, and their near allies, the *Elateridæ*, or Fireflies, are sometimes used as ornaments, for which purpose the former are frequently employed even in Europe. The Diamond Beetles, which are large *Curculionidæ*, or Weevils, form magnificent objects for the microscope.

But much more might be done to make insects useful. The Cockchafer, one of the most abundant and destructive of all our British Beetles, might be made an insect of great commercial value,—as has been proved, though more in the way of experiment than with any practical result; for in this age of the world, capitalists prefer to invest their money rather on products of established value than in promoting new conquests from the kingdom of nature. Cockchafers form a very fattening food for fowls; they will yield oil, which burns with a bright flame; grease can be obtained from them which is useful for greasing carriage wheels; and it is even said that a dye can be obtained from them. There is no doubt that with a little patience and experiment, beetles might be made very useful to man in a great variety of ways in which we have at present no idea.

But we cannot turn insects to any practical value without taking up the study of Entomology seriously, and not simply as an amusement; for many insects that look very much alike to inexperienced eyes, are really very different indeed.

We may find some plant in our fields or gardens suffering severely from the attacks of insects, and swarming with some species of insect which, so far from being the real depredator, is busily engaged in diminishing its numbers. In such a case, any one unacquainted with Entomology would probably devote all his energies to destroying his benefactors, while the real authors of the mischief might very possibly escape scot-free.

It is quite certain, too, that if any one wished to use an insect for any special medical or commercial purpose, and had only a general idea of what it was like, he would be almost certain to pitch upon something else, which might happen to possess very different properties indeed from the insect he was really in quest of.

The usefulness of a knowledge of Entomology was ludicrously illustrated by the Colorado Potato-Beetle panic some years ago, when every one who found an insect among his potatoes at once wrote to the local newspaper to announce the appearance of the dreaded enemy. Some of these blunders were compara- tively venial, as when the common Ladybirds (beetles somewhat similar in general appearance) happened to do duty as Colorado Potato-Beetles ; but others were monstrously wide of the mark, as when the larva of the Death's Head Hawk-Moth (*Acherontia Atropos*), a great yellow-striped caterpillar six inches long, was the supposed beetle. An even more absurd case once fell under my own notice. An Irishman went down on the quay, and found a crustaceous animal, closely allied to the wood-lice, but much larger, which is very common on the sea-shore, running along a tow-rope. He at once seized it, exclaiming, "Here's the Colorado Potato-Beetle just landing from America!" and took it home with him, when he immediately wrote to the paper ; and there was quite a sensation in the town for a day or two, until the mistake was discovered and exposed.

BRITISH BUTTERFLIES AND MOTHS.

E

BRITISH BUTTERFLIES AND MOTHS.

ESIRE to form a collection of butterflies and moths inspires almost every boy who has ever lived in the country. They are beautiful insects, and are easily to be obtained and preserved; but you cannot do so successfully unless you know how to set about it. Firstly, then, never touch your specimens with your fingers, without care; for the down on the body and the scales on the wings are easily rubbed off, and the specimens are then spoiled; besides, the wings themselves are very fragile, and easily broken.

Butterflies, and indeed all insects which are captured on the

wing, are generally collected with the aid of a net. I myself
prefer a common ring-net, which is made of a
jointed iron ring which screws on to the end of
a walking-stick, and can be folded up and put
in the pocket when not in use.   The ring is
about 9 inches across, and to this is attached
a strip of stout green chintz, on which is sewn a
net of green gauze, about 18 inches in length.
Such a net may be bought of any of the dealers
in objects of natural history for about 4s. 6d.
The "umbrella-net" is formed of similar materials,
but is mounted on a whalebone ring instead of an
iron one, and slides up and down a stick, being
covered by a common umbrella case when not
in use.   But this form of net is much more
costly than the other, and the stick is too short
for many purposes.   It is, however, very easy to
manufacture a net for yourself out of a ring of
flexible twigs, a piece of green gauze, and a light
Y-shaped sapling, between the arms of which
the net is fixed.   The net should always be
transparent, and should contain no corners ;
green is the best colour, as it harmonizes with
the colour of grass and trees.

You will find butterflies and day-flying moths
in gardens, fields, and woods ; and nocturnal
species may be found at rest on shady walls,
tree-trunks, or in outhouses, or may be disturbed from their lurking places by beating a
hedge to windward.   In the evening, many
moths may be captured flying over flowers, or
may be attracted into a room by a light placed at
Ring-net.
an open window, while others may be obtained by "sugaring,"
that is, painting the trunks of trees with a mixture of sugar
and beer, flavoured with a few drops of rum.   The patches of
"sugar" must then be visited after dark with a lantern, when
moths will frequently be found regaling themselves upon the
sweet mixture.

Sluggish moths and small moths may generally be safely
carried home alive in pill-boxes, taking care not to mix full and
empty ones, and only to put one specimen in each box ; but
butterflies and active moths must be killed and pinned at once.
Butterflies and small moths may be carefully nipped below the
wings, taking care not to damage them.   Stout-bodied moths

cannot be thus killed without being completely spoiled; for killing insects in the field, a glass jar about the size of a jam pot, and stopped with a bung, is usually employed, charged with a strong poison, which may be purchased ready-made where you buy your net. But a piece of blotting paper soaked in benzole is used by some Coleopterists, and might answer equally well for *Lepidoptera;* besides, it would not injure their colours, as some of the chemicals employed are liable to do. A bell-glass or a deep glass jar will be found more convenient for killing insects brought home alive; and if a small hole be made in the lid of the pill-boxes, they may be dropped into the killing jar, without being opened till the enclosed moth is dead.

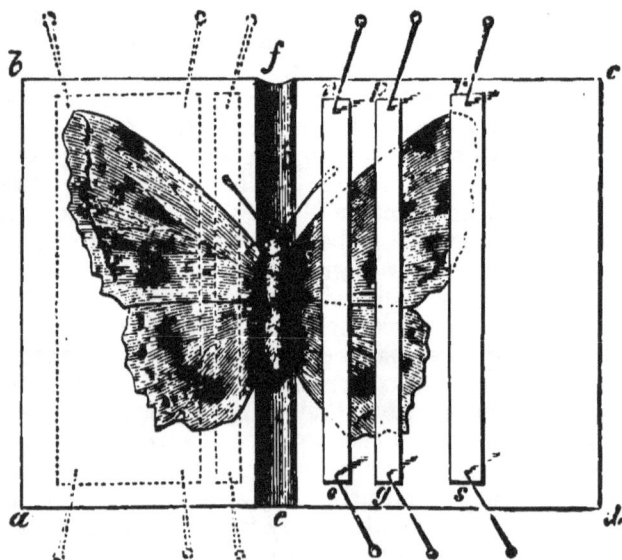

Setting-board.

Common pins are too thick and clumsy to be used for pinning insects; those used for insects are long and slender, and may be bought of any dealer in objects of natural history. In order to set insects, you require setting-boards, which are made of flat pieces of deal, of any length you please, and from one to six inches in width. There is a groove in the middle, of any convenient depth, but it must be uniform in all your boards, and should be deep enough to keep the insect well off the paper of the drawer, when set.   The groove, as well as the sides of the board (which may be either flat or bevelled off), must be covered with cork.   Pin your insect through the centre of the thorax, stick it in the groove, and arrange the wings on each side of the board in a natural position, and fix them down with strips of card-board (as shown in our illustration), and leave them till stiff. Then arrange them in a tight-fitting corked box, placing them in rows, and as nearly in order as you can, putting the name of the genus above and the name of the species below your series of each species.

A cabinet being an expensive article, you had better keep your collection in boxes at first.   Tightly fitting boxes, like backgammon boards, but rather deeper, and lined. with cork top and bottom, are the best; and smaller boxes of a similar kind are necessary, to carry about in the pocket.   If you have a turn for mechanics, you can perhaps amuse yourself by making boxes for yourself, lining them with sliced bottle-corks, if you have nothing else handy, and pasting clean white paper over the corks to make the box look neater, and to show off the insects better.   It is a good plan to brush over the paper with a little carbolic acid and water (just so weak as to leave no stain), and then let it dry before using it.   Insects must be kept in the dark, for light bleaches them, and a little camphor must be kept in the box, and replenished as often as necessary, or they will soon be devoured by mites.   The carbolic acid is an additional safeguard.

In rearing caterpillars, avoid touching them with the fingers, and keep them plentifully supplied with fresh food, which should not be gathered when wet, and the old food should be carefully removed.   In collecting perfect insects, never catch more specimens than you want for your own collection, or for your friends ; and do not keep any damaged specimen, unless it is a rarity which you are not likely to be able to replace.   It is true that most insects are generally abundant where they occur ; but many are confined to certain localities, and it is a pity to destroy them wantonly, especially when you perhaps run the risk of materially reducing the numbers of a local species.

Although I cannot here attempt to give such a complete out-line of British Lepidoptera as is included in my larger work on European Butterflies and Moths, yet I will now attempt to give a brief introductory sketch of the subject, which may be useful to beginners.

Butterflies and Moths belong to the Order *Lepidoptera* or Scale-winged Insects ; they pass through four well-marked

Melanargia Galathea.

Argynnis Paphia.

Polyommatus Corydon.

[Face p. 70.

stages—egg; larva, or caterpillar; pupa, or chrysalis; and imago, or perfect insect. These changes are called metamorphoses, or transformations; and they are complete in *Lepidoptera*, which means that the four stages are all sharply separated from each other. A caterpillar has six horny legs in the front of its body, and from four to ten additional fleshy legs, called prolegs, on the hinder segments of its body. The two last of these are called claspers. A butterfly or moth has only six legs, corresponding to the horny legs of the larva; but occasionally either the first or the last pair is aborted, especially in the males. They have four wings, clothed with scales, and imbibe their food through a proboscis, although caterpillars have mandibles, and bite their food. The first five families of *Lepidoptera* are called *Rhopalocera* (Knob-Horns), because their antennæ, or feelers, are more or less thickened into a knob at the tip. The butterflies, of which we have sixty-five different kinds in England, fall into this division. The moths are called *Heterocera* (or Various-Horned), because their antennæ are of various shapes, sometimes tapering gradually to the tip, sometimes of uniform thickness throughout, sometimes thickest in the middle, and sometimes more or less comblike or feathery, when they are said to be pectinated.

The five groups, or families, into which butterflies are divided, are called *Nymphalidæ, Erycinidæ, Lycænidæ, Papilionidæ,* and *Hesperiidæ.* The *Nymphalidæ* have the forelegs rudimentary in both sexes, and the pupa is suspended by the tail. It is divided into two subfamilies, *Satyrinæ* and *Nymphalinæ.* The *Satyrinæ* are brown butterflies, more or less marked with tawny, and always with a round spot, either in a pale ring, or with a white dot in the middle, at the tip of the fore wings, and often others near the borders of the hind wings. They vary from an inch and a half to two inches in expanse, and many are very common. The Meadow Brown (*Epinephile Janira*), which is brown, with a tawny patch on the fore wings of the female, abounds in every field; the Ringlet (*E. Hyperanthus*), which is blackish brown, with a row of eyes on all the wings beneath, is common in woods; and the Grayling (*Hipparchia Semele*), which is brown, with tawny markings, and two eyes on the fore wings, is common in waste places. The Speckled Wood, or Wood Argus (*Satyrus Ægeria*), is brown, with yellowish-white spots towards the margins, and is found in woods in spring; while the Wall-Brown (*S. Megæra*) is a handsome brown and tawny butterfly, common in lanes, etc., and fond of sunning itself on walls. The Marbled White (*Melanargia Galathea*) is a conspicuous black and white butterfly, which is very local, though abundant where it occurs; and the Small Heath Butterfly (*Cœnonympha*

Wall-Brown (*Satyrus Megæra*).

Pamphilus) is a sandy-coloured butterfly, smaller than any we have mentioned, which is very common in open places.

To the *Nymphalinæ* belong many of our handsomest and most conspicuous butterflies. The Fritillaries of the genus *Argynnis* vary in expanse from one and a half to three inches. They are of various shades of fulvous, with black spots or markings on the upper surface, and the under side of the hind wings is always spotted or streaked with silvery white. The two smallest species (*Argynnis Selene* and *Euphrosyne*) are common in woods in spring; but the larger species appear in summer, when the Dark Green Fritillary (*Argynnis Aglaia*) frequents heaths. It derives

Small Tortoiseshell (*Vanessa Urticæ*).

its name from the green colour of the under surface of the hind wings, which are likewise marked with many silver spots; but the most beautiful butterfly of this group is the Silver-washed Fritillary (*A. Paphia*), which is streaked with silver on the under surface of the hind wings, instead of being spotted. It is common in woods, but is not always easy to catch. The spiny caterpillars of the species of *Argynnis* feed on violets. There is another genus of Fritillaries (*Melitæa*) which includes three black and tawny species, all very local. They are not spotted with silver, and their larvæ feed on plantain.

There are three very common and beautiful butterflies the larvæ of which feed on nettle. These are the Small Tortoiseshell (*Vanessa Urticæ*), the Peacock (*V. Io*), and the Red Admiral (*Pyrameis Atalanta*). In the two first each wing has a slight projection, giving them an angular appearance; but that on the fore wings of the Red Admiral is less acute, and there is none

Peacock Butterfly (*Vanessa Io*).

on the hind wings. The Small Tortoiseshell is bright reddish, with black spots on the fore wings, and the basal or inner half of the hind wings black; the borders of all the wings are dusky, with a row of small blue spots. It expands two inches, or a little over. The Peacock is of a dull red colour, with a large black space on the hind wings, partly bordered with buff, and filled up with blue markings; the fore wings are spotted with black and yellowish on the costa, or front edge; and there is a roundish composite yellow, black, and blue spot towards the tip. The Red Admiral is black, with a red band on the fore wings and a red border on the hind wings, and some white spots towards the tip of the fore wings. Both these butterflies are rather larger than the Tortoiseshell. The Painted Lady (*Pyrameis Cardui*) is of a pale salmon colour, with black markings, and some white spots towards the tip of the fore wings. Its wings are less angulated than even in the Red Admiral. Its larva feeds on thistle, and the butterfly is much commoner in some years than others. In some years it is extraordinarily abundant, and migrates in vast swarms from one part of the country to another. The Comma Butterfly (*Vanessa C. Album*) is a local insect generally found flying along hedges. It is fulvous, with dark markings, and is about the size of the Small Tortoiseshell, but may be recognised at once by its very jagged wings. The White Admiral (*Limenitis Sibylla*) measures over two inches across the wings, and is black, with a white band, more perfect on the hind wings than on the fore wings. It is a local insect, found in woods in the South of England; and the same may be said of the Purple Emperor (*Apatura Iris*), one of the finest of our British Butterflies. It is brown, banded with white, nearly as in *L. Sibylla*, but the male is suffused with the richest purple, and soars over the tops of the trees, whereas the White Admiral has a lower and more sailing flight. The Purple Emperor measures about three inches across the wings.

The family *Erycinidæ* only includes one European species, called the Duke of Burgundy Fritillary (*Nemeobius Lucina*). It measures a little more than an inch across the wings, which are

Duke of Burgundy Fritillary (*Nemeobius Lucina*).

black, with rows of yellowish spots. It is found in woods in May and June, but is not generally common. The female has six perfect legs, but the first pair are imperfectly developed in the male. In the three following families both sexes have six legs.

The *Lycænidæ* are a family of small butterflies, all of which are brown, coppery red, or blue. The Hairstreaks, belonging to the two genera *Zephyrus* and *Thecla*, usually have pale lines on the under side of the wings, and there is a delicate projection from the middle of the hind wings, which is called a tail. They are all rather scarce, except two species. The Purple Hairstreak (*Zephyrus Quercus*), is dull blue in the male, and brown, with a bright purple blotch at the base of the fore wings, in the female.

Green Hairstreak (*Thecla Rubi*).

It expands nearly an inch and a half, and is found in oak woods in July. The Green Hairstreak (*Thecla Rubi*), on the contrary, is common in bushy places in spring. It is rather smaller than the Purple Hairstreak, and is brown above and green below, and there is no tail on the hind wings.

The Small Copper Butterfly (*Lycæna Phlæas*) measures rather

Small Copper (*Lycæna Phlæas*).

more than an inch across the fore wings, which are coppery red, with black spots ; the hind wings are blackish, with a coppery red border. The species of *Polyommatus* are seldom much larger, and are either blue or brown in both sexes, or the males are blue and the females brown. Most of the species are common where they are found ; but the Common Blue (*P. Icarus*) is the only one which, like the Small Copper, is abundant everywhere. It expands about an inch and a quarter. The male is blue, with whitish fringes to the wings, and the female is brown, more or less suffused with blue, with reddish spots towards the borders of the wings.

The *Papilionidæ* include all our white and yellow butterflies. This family is divided into two sub-families called *Pierinæ* and *Papilioninæ*. The last is only represented in Britain by the Swallow-Tail Butterfly (*Papilio Machaon*), a great black and yellow butterfly, with a large red spot near the inner angle of the hind wings, which are also furnished with a long tail in the middle. This butterfly is confined to the fens of the Eastern counties.

On the other hand, the *Pierinæ* include several of our commonest butterflies. Three white butterflies—the Large White (*Pieris Brassicæ*), the Small White (*P. Rapæ*), and the Green-veined White (*P. Napi*)—are common in every garden through-

Clouded Yellow (*Colias Edusa*).

out the summer, and their caterpillars feed on cabbages, etc. Another species, often seen in spring, is the Orange-Tip (*Euchloe Cardamines*), which is white, with the hind wings mottled with bright green beneath, and a bright orange spot towards the tip of the fore wings in the male. The Brimstone Butterfly (*Gonepteryx Rhamni*), which is of the colour expressed by its name, is common in woods, and is one of the earliest butterflies in the year to appear. It has an angular projection on each of its wings. It expands about two inches and a half. The Clouded Yellow Butterflies (*Colia Edusa* and *Hyale*) have orange or yellow wings with dark borders, and no projection on the hind margins. They are smaller than the Brimstone and are much commoner in some years than others, though, on the whole, they are far more abundant now than they were forty or fifty years ago.

The *Hesperidæ* are small butterflies with large heads, thick bodies, and the antennæ, which are sometimes hooked, inserted widely apart instead of close together. The Grizzled Skipper (*Hesperia Malvæ*) is dark brown, tessellated with white spots ; and the Large Skipper (*Pamphila Sylvanus*) is greenish brown with fulvous markings, and a broad black streak at the base of

Small Skipper (*Thymelicus Thaumas*).

the fore wings in the male. The Small Skipper (*Thymelicus Thaumas*) is fulvous, with blackish borders, and a black streak on the fore wings in the male. None of these species measure much more than an inch in expanse ; and they are all very common in woods, etc.

The moths are divided into numerous families, and these are sometimes arranged in larger sections, called *Sphinges, Bombyces, Noctuæ, Geometræ, Pyrales, Tortrices, Tineæ, Pterophori*, and *Alucita*. But the first two sections are generally used to include very discordant groups, and had better be dropped.

The first four of these nine groups are often included with the butterflies under the general term *Macro-Lepidoptera*, while the last five groups of moths, which include small species only, are called *Micro-Lepidoptera*.

We will now speak of the principal families included under the terms Sphinges and Bombyces. The Sphinges are usually employed to denote the families *Sphingidæ, Ægeriidæ*, and *Zygænidæ ;* but the second of these is widely different from the first, while the third should more properly be included with the Bombyces.

The *Sphingidæ*, or Hawk-Moths, include our largest moths. They have stout bodies, long narrow wings, more or less pointed, and their flight is generally very rapid. Our smallest species are the Humming-Bird Hawk-Moth (*Macroglossa Stellatarum*) and the Broad-Bordered and Narrow-Bordered Bee Hawk-Moths (*Hemaris Fuciformis* and *Bombyliformis*). The Humming Bird Hawk-Moth has brown fore wings, and brownish-red hind wings, and may be seen hovering over flowers during the day or at dusk during a great part of the summer. The Bee Hawks may be known by their transparent wings, with brown borders, and are found flying over flowers in or near woods, in spring, during the day. These three species have a spreading tuft of hairs at the

Narrow-Bordered Bee Hawk-Moth (*Hemaris Bombyliformis*).

end of the body, which we do not meet with in the remaining *Sphingidæ*. These all fly in the evening, or at night.

The Death's Head Hawk-Moth (*Acherontia Atropos*), which sometimes measures nearly six inches across the wings, is not only the largest British moth, but one of our largest British insects. It has brown fore wings, and pale yellow black-banded hind wings, and there is a yellow pattern on the back of the thorax which has some distant resemblance to the shape of a skull. It has the power of uttering a very audible squeak, and is fond of honey, sometimes entering bee-hives and committing considerable damage. Its enormous yellowish larva is frequently found feeding on potato-leaves in autumn, and, like most of the *Sphingidæ*, it has a long fleshy horn on the back of the last segment but one. Although not rare, it is never sufficiently common to do any real injury to our potato crops, though they frequently suffer from the attacks of smaller but more abundant insects.

Caterpillar of Privet Hawk-Moth (*Sphinx Ligustri*).

The Privet Hawk-Moth (*Sphinx Ligustri*) is a commoner

insect, rather smaller than the Death's Head. It measures about four inches across the fore wings, which are light brown, with darker markings ; the hind wings are pale pink, with transverse black bands. Its green caterpillar feeds on privet and lilac.

Eyed Hawk-Moth (*Smerinthus Ocellatus*).

In the genus *Smerinthus*, the hind margins of the wings are dentated ; and one of our prettiest moths is the Eyed Hawk-Moth (*S. Ocellatus*). The fore wings are of a reddish grey, with lighter and darker markings, and the hind wings are rose-colour, with a large round black spot near the hinder angle, enclosing a blue ring. It expands about three inches.

Ægeria Apiformis.

The *Ægeriidæ* are not numerous ; and their larvæ are whitish grubs which feed inside the stems of trees and plants. Their

wings are more or less transparent, and they resemble hymenop-
terous insects rather than moths.   The largest and commonest is
Ægeria Apiformis, a black, yellow-belted sluggish insect, which
is often found resting on the trunks of poplars in early summer.
It so much resembles a large wasp that when I first saw it alive, I
felt very nervous about touching it with my fingers, although I
knew perfectly well that it was only a moth.   The species of
the genus *Trochilium* are small moths which only expand about
an inch.   They have transparent wings, with blackish borders,
sometimes tinged with yellow or red ; and long bodies, tufted at
the tip, and marked with one or more red or yellow belts.
The commonest species is *T. Tipuliforme*, which is found among
currant bushes, on the pith of which the larva feeds ; the moth
has three yellow belts on the abdomen.

The *Zygænidæ* are small moths with long wings, and short

Five-Spot Burnet Moth (*Zygæna Trifolii*).

bodies ; they fly by day.   The Burnet Moths have steel-blue fore
wings, streaked with red, or marked with five or six red spots.   The
commonest species are *Z. Filipendulæ* (with six) and *Z. Trifolii*
(with five).   Their antennæ are much thickened before the tip,
which is not the case in the Green Foresters, of which we have
three local species, all very much alike, with green fore wings
and dark-brown hind wings.   They are a little smaller than
the Burnets, only measuring about an inch across the wings.
Procris Statices, which frequents meadows, like most of the
Zygænidæ, is the commonest species.

Turning now to the *Bombyces*, we find that the family *Arctiidæ*,
or   Tiger-Moths,   contains   the   handsomest   species.   The
Cinnabar   Moth   (*Callimorpha Jacobææ*)   is very   similar   to   a
Burnet, and is almost the same size, but has much broader wings,
and its antennæ are not thickened.   It is brown, with red hind
wings, and red streaks and spots on the fore wings.   The Tiger
Moth (*Arctia Caja*) is very common in gardens, and its shaggy
brown and reddish caterpillar, which rolls itself up into a ball

when touched, feeds on a variety of low plants. The moth expands about two inches and a half, and has brown fore wings, with interlacing white markings, and red hind wings, with blue-black spots. The White and Buff Ermines (*Spilosoma Menthastri*

Tiger-Moth (*Arctia Caja*).

and *Lubricipeda*) are also very common in gardens; they are white or yellowish white, with numerous black dots on their wings; they measure about an inch and a half in expanse.

The *Lithosiidæ* have rather narrow fore wings and broad hind wings. Their bodies are short and slender, and they are generally pale coloured, often with a yellowish stripe along the front of the fore wings. They fold their fore wings flatly over each other, and sham death when alarmed. They rarely measure more than an inch and a half in expanse, and are sometimes more or less spotted; one species (*Gnophria Rubricollis*) is black, with a red band on the neck, and the body tipped with red.

The *Liparidæ* are broad-winged moths with rather short

Brown-Tail Moth (*Porthesia Chrysorrhœa*).

bodies, and the males have pectinated antennæ. Many of the species are white, such as the Gold-Tail and Brown-Tail Moths

F

(*Porthesia Auriflua* and *Chrysorrhœa*), which are provided with a tuft of down at the extremity of the body, which the female uses to cover her eggs.   The moths are common on hedges at dusk in summer.   They expand about an inch and a half across the wings.   The White Satin Moth (*Stilpnotia Salicis*), another very common species, is rather larger, and is pure white, with no coloured tuft at the extremity of the body, and no black spot towards the hinder angle of the fore wings, as is generally the case in *Porthesia*.

Commoner still is the Vapourer Moth (*Orgyia Antiqua*), which

Vapourer Moth (Male).
(*Orgyia Antiqua*).

Vapourer Moth (Female).
(*Orgyia Antiqua*).

measures about an inch across its reddish-brown fore wings, which have a white spot near the hinder angle ; the hind wings are paler.   The male may be seen fluttering about in the day-time everywhere where trees or bushes grow ; for its tufted larva is not particular about its food, and will eat even laurel. The female moth has rudimentary wings.

The *Psychidæ* are small black moths, with broad rounded wings, expanding an inch or less.   The males fly among the grass by day, but the females are apterous.   The larvæ form cases for themselves, like those of caddis-worms.

The *Notodontidæ* are large whitish or brownish moths, with rather long bodies, more or less tufted at the extremity, and the males have pectinated antennæ.   One of the largest and com-monest species is the Puss Moth (*Cerura Vinula*), which measures nearly three inches across the wings, which are greyish-white, with zigzag black lines on the fore wings.   The green larva feeds on poplar and willow, and its last pair of prolegs are converted into two long tail-like appendages which enclose retractile threads.   Another well-known representative of this family is the Buff Tip (*Phalera Bucephala*), which is rather

smaller than the Puss Moth. The fore wings are varied with
grey and brown, and there is a large pale yellowish spot at the
tips; the hind wings and abdomen are yellowish-white. Its

Puss Moth (*Cerura Vinula*).

downy larva, striped alternately with black and yellow, is very
common on various trees.

Our only British representative of the family *Saturniidæ* is the
Emperor Moth (*Saturnia Pavonia-minor*). It is a broad-
winged moth, measuring nearly three inches in expanse. The
fore wings of the male are varied with dark grey, and the hind
wings are reddish-yellow; in the female all the wings are pale
grey. In both sexes there is a large black eye in the middle of
the wing, with a whitish crescent in the middle, and surrounded
by yellow and black rings. The green larva, with red warts
set with short black hairs, is common on heath.

The *Lasiocampidæ* comprise a number of large brown, reddish,
or yellowish moths, with thick hairy bodies. The larvæ are
also hairy; and as in the *Saturniidæ*, the pupa is always
enclosed in a tough silken crown. The three commonest species
are the Oak Eggar, the Drinker, and the Lackey. The Oak
Eggar (*Lasiocampa Quercus*) is about the size of the Emperor
Moth; and as is the case in most butterflies and moths, the
female is rather larger than the male. The male is chestnut
coloured, with a yellowish band, suffused on the outside, beyond
the middle, and the female is yellowish. The larva feeds on
oak, heath, etc. The male flies by day so rapidly that it is
impossible to run it down; but the moth is easily bred from the
larva; and a newly emerged female, carried in a box in the
collector's pocket, will entice any males within reach which may

happen to be in the neighbourhood. The Drinker Moth (*Odonestis Potatoria*) is reddish-brown in the male, and yellowish in the female, with some dark lines on the wings, and one or two small white spots towards the costa, or front edge of the fore wings ; it is generally considerably smaller than the Oak Eggar. Its brown hairy larva, with reddish sides, and a tuft of black hairs behind the head, and another near the tail, is common everywhere on grass. The Lackey Moth (*Clisiocampa Neustria*) is a much smaller insect, not measuring more than an inch and a half across the wings, which are yellowish or reddish, with two dark transverse stripes on the fore wings. The eggs are laid in a close ring like a bracelet round the slender branch of a tree, and the caterpillars are striped with black, blue, and reddish orange, with a longitudinal white stripe on the back. They feed on fruit-trees, etc., and spin themselves a common web, to which they retire for the night ; but their large nests are so conspicuous that they can easily be seen and destroyed.

Xyleutes Cossus, the Goat Moth, the commonest species of the small family *Zeuzeridæ*, is a large, greyish-brown, heavy-looking moth, with thick body, and broad wings, expanding about three inches. Its great reddish caterpillar lives in long galleries which it eats in the wood of willows and other trees, and is three years in arriving at maturity.

The *Hepialidæ*, the last family of the *Bombyces* which we shall notice, may be illustrated by the Ghost Moth (*Hepialus*

Ghost Moth (*Hepialus Humuli*), Female.

Humuli), which has a peculiar hovering flight, and is common in meadows on summer evenings. The male is white above and brown below, and therefore keeps appearing and disappearing in a manner from which it has probably derived its name. The female is reddish-yellow, with red markings on the fore wings. The yellowish-white larva feeds on the roots of grass. The other species of *Hepialus* are smaller insects, dark coloured, with pale spots or markings on the fore wings, and some fly rapidly near the ground, while others have a hovering flight like the Ghost Moth.

The *Noctuæ* form a large and compact group of moths. They are seldom brightly coloured ; but the fore wings are generally of some dark colour, and the hind wings are brown or white, often iridescent, and generally without markings ; it is rare for the markings of the fore wings to be continued on the hind wings in this group. The body is moderately stout ; the thorax is generally covered with hair, and sometimes crested, and the abdomen is rather long, often extending beyond the hind wings ; the antennæ are generally simple.

We have mentioned all the families of the Butterflies and Sphinges, and nearly all those of the Bombyces ; but in the Noctuæ and following groups we can only mention a small number of representative species. The first which we have

Peach-Blossom Moth (*Thyatira Batis*).

selected is the Peach-Blossom Moth (*Thyatira Batis*), belonging to the family *Cymatophoridæ*. It is a very pretty moth, with dark-brown fore wings and grey hind wings, the former with several large spots of a very delicate pink and white. It ex-pands an inch and a half, which may be considered about the average size of the *Noctuæ*. The Dark Arches (*Xylophasia Polyodon*), and the Cabbage Moth (*Mamestra Brassicæ*), are both very common in gardens. The former is a brown moth, with some dark marks on the fore wings, near the hind margin of which is a pale line, shaped like a W. The moth expands two inches, and has a long body, tufted at the extremity. The Cabbage Moth is smaller, with a shorter body ; the fore wings are blackish-brown, with a pale W (a common mark in several genera of *Noctuæ*) near the edge, and with a pale mark in the middle. The caterpillar is very destructive to cabbages. Both these moths belong to the family *Apamidæ*.

The family *Agrotidæ* includes the most typical moths of the

Noctuæ; their caterpillars feed on the roots of plants, and are often very destructive. Most of the species of *Agrotis* have brown fore wings, and brown or white hind wings; but the genus *Triphæna,* including the Yellow Underwings, has brown

Yellow Underwing (*Triphæna Orbona*).

fore wings, and yellow hind wings, with a black border. These moths may often be met with by day in dark corners, among long grass, or strawberry beds, etc.

The species of *Cucullia,* belonging to the family *Xylinidæ,* are brown or grey moths, called "Sharks" by collectors. They have rather long, narrow, and often pointed wings, and long bodies. They expand nearly two inches, and fly over flowers in the evening; but their long, simple antennæ, which are not thickened in the middle, will at once prevent any danger of their being mistaken for small *Sphingidæ,* to which they have some outward resemblance.

The Gamma Moth (*Plusia Gamma*), belonging to the family

Gamma Moth (*Plusia Gamma*).

Plusidæ, is one of our commonest *Noctuæ,* and is, perhaps, the most frequently observed of any, as it flies over flowers by day as well as at dusk. It has violet-grey fore wings, in the middle of which is a white mark shaped like a Y or the Greek letter Gamma.

Plusia Chrysitis (the Burnished Brass Moth) flies in the evening. It is about the same size as the Gamma Moth, but the fore wings are nearly covered by two large brassy-green blotches, which are generally connected.

Mormo Maura (the Old Lady), belonging to the family

Old Lady (*Mormo Maura*).

Toxocampidæ, is a large dark-brown moth, which is common in gardens in the evening, and often flies into houses when the windows are open.

The Red Underwing (*Catocala Nupta*), belonging to the family *Catocalidæ*, measures about three inches across the fore wings, which are varied with grey. The hind wings are red, with a black border and a black band across the middle. It may often be found sitting on the trunks of trees in the daytime, to which the colour of its fore wings assimilates it, its red hind wings being covered over, and invisible. It has then a triangular form, which is still more noticeable in the *Deltoidæ*, which differ from the other *Noctuæ* by their slender bodies, and by the palpi (two organs projecting from the head between the antennæ) being so much developed as to resemble a kind of beak. The commonest species is the Snout (*Hypena Proboscidalis*), which is found among nettles; it measures an inch and a half across the fore wings, which are brown, with two darker stripes. In the genus *Herminia* the males are provided with a curious tuft of hair on the front legs.

Herminia Tarsipennalis.

The *Geometridæ* are a large group of broad-winged, slender-bodied moths, generally of gay colours, and with the markings of the hind wings more or less similar to those on the fore wings. Their caterpillars have only ten legs—six in front, and four behind—and are therefore obliged to walk in a very peculiar fashion, arching their backs at every movement. These moths are readily disturbed from their hiding-places during the day, and are not very strong on the wing; some few fly naturally by day, but the greater number fly at dusk.

Several species are green, with whitish lines across the wings. The commonest is *Hemithea Thymiaria*, which expands rather more than an inch, and may be beaten out of hedges in summer; it differs from most of the allied species by the hind wings being angulated. The genus *Acidalia* includes a great number of small white moths, with dark lines on the wings. They seldom expand much more than an inch, and many are smaller; they

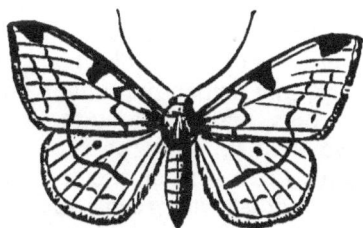

Brimstone Moth (*Rumia Cratægata*).

are generally called "Waves" by collectors. The Magpie Moth (*Abraxas Grossulariata*) is common everywhere in gardens, where its caterpillar feeds on gooseberry and currant bushes. It is white, with rows of black spots on the wings, and orange bands at the base and in the middle of the fore wings. The

Swallow-tail Moth (*Urapterya Sambucaria*) expands nearly two inches. It is pale yellow, with two dark lines on the fore wings and one on the hind wings ; the hind wings project into a short angular tail. The Brimstone Moth (*Rumia Cratægata*) is a smaller insect, very common about hedges, for its larva feeds on hawthorn. It measures about an inch and a half across the wings, which are sulphur yellow, with some rust-coloured spots towards the costa. Some of the *Geometridæ* belonging to the

Autumn Moth (Male.)
(*Hybernia Defoliaria.*)

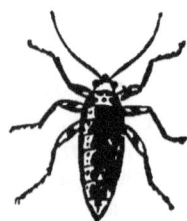

Autumn Moth (Female.)
(*Hybernia Defoliaria*).

genera *Hybernia* and *Cheimatobia* appear only in winter; and these have all apterous females. The fore wings of the males are yellowish or brown, with darker markings, and the hind wings are paler. There is another section which differs from the other *Geometridæ* in having very stout bodies. Most of these appear in early spring, and many of them have also apterous females ; but the Pepper-and-Salt Moth (*Amphidasis Betularia*) is an exception. It is white, speckled with black, and expands nearly two inches ; it appears in May. *Odezia Atrata* (the Chimney Sweep) is a smoky black moth, with white tips to the fore wings, and expanding about an inch ; the body is slender. The Carpets, which belong to the genera *Coremia*, *Larentia*, etc., have whitish, grey, or brown fore wings, with darker zigzag patterns, the centre of the fore wings being often banded with brown, reddish, or green ; they generally expand at least an inch. The moths of the genus *Eupithecia* (called popularly "Pugs") are usually of very dull colours, and of small size, only the largest species attaining the expanse of an inch.

The typical *Pyralidæ* are small moths, with broad rounded wings, slender bodies, and long slender legs. Some are found in houses, like the Tabby Moth (*Aglossa Pinguinalis*), which is brown, and expands about an inch and a half; others are found in woods, like the black, white-spotted *Ennychia Octomaculata,* or frequent dry places, like the red, yellow-spotted *Pyrausta Purpuralis;* others, like the China Marks belonging to the genera *Cataclysta* and *Hydrocampa,* are found among reeds; these latter

Ennychia Octomaculata.

are white, with black and sometimes with yellowish lines on the wings. The *Botydæ* are rather larger and longer-winged insects than the true *Pyralidæ,* and are generally of a yellowish colour; one species, however (the Small Magpie, *Botys Urticata*), which is common among nettles, is black and white.

The *Crambidæ* are small moths, with slender bodies, and long palpi. The fore wings are narrow, and the hind wings ample, but fold into a very small compass when at rest. Many species of *Crambus,* with pearly white, yellowish, or brown fore wings streaked with white, and brown or whitish hind wings, are common in meadows.

The *Tortricidæ,* or Bell Moths, have broad ample wings, and broad, rather short, truncated fore wings, so that they somewhat resemble a bell in shape when at rest. They are all small moths, rarely expanding more than three-quarters of an inch. Many of their larvæ live in rolled-up leaves; others live in the heads of composite flowers, or in the interior of fruits. The Green Oak Moth (*Tortrix Viridana*), with green fore wings and brown hind wings, may be beaten in abundance from any oak tree in summer. The larva of *Penthina Pruniana,* the brown moth

Penthina Pruniana.

with paler markings which we have figured, feeds on sloe; but the larvæ of two other species of this group (*Carpocapsa Pomonella* and *Funebrana*) live in the interior of apples and plums. Two species of *Xanthosetia,* with rather narrower wings than the majority of the *Tortricidæ* (*X. Hamana* and *Zægana*), are common on the heads of thistles and other composite flowers; the fore wings are yellow, with rust-coloured markings, and the hind wings are light brown.

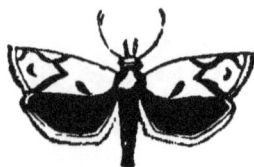

Xanthosetia Hamana.

The *Tineæ* are a very large group of small moths, including nearly a third of the British *Lepidoptera.* They may generally be known by their rather long and narrow wings, with very long fringes. They are divided into many families, of which we can only mention a few. The *Tineidæ* include the bulk of the Clothes-Moths, which give so much trouble to our housekeepers, and the *Adelidæ* may be known by their green or brown wings, and their

Nemophora Swammerdamella.

very long antennæ. The *Hyponomeutidæ* include the Small Ermine Moths, which have white or grey fore wings, with several rows of black dots. Their larvæ are gregarious, spinning a common web, and frequently stripping our hedges and apple-trees of

Phibalocera Quercana.

their leaves. *Phibalocera Quercana* is a very pretty species belonging to the great family *Gelechidæ.* The fore wings are

reddish grey, with yellow spots, and the hind wings are whitish ;
the larva feeds on oak, beech, etc., in a web on the under surface
of the leaves. The *Coleophoridæ* include a number of small, long-
winged species, many of the larvæ of which live in cases, something like those formed by the larvæ of *Psychidæ.* *C. Vibicella,*

Coleophora Vibicella.

which we have figured, is not a very common species ; it is
bright ochre-yellow, with silvery-white streaks on the fore wings.
The larvæ of the *Tineæ* are very various in their habits, some
feeding between united leaves, others forming galls, and a great
number feeding in the substance of the leaves, and forming
blotches or galleries by which their presence can be easily detected.   The smallest species belong to the family *Nepticulidæ,*
the smallest of all being *Nepticula Microtheriella,* the larva of
which feeds in nut-leaves.   Many of these very small species are
very beautiful, being of rich dark colours, relieved by metallic
spots.
    The Plume Moths, of the family *Pterophoridæ,* are delicately

Plume Moth (*Pterophorus Lithodactylus*).

formed moths, with long bodies, long slender legs, and rather
narrow wings, the fore wings being split up (except in *Agdistis
Bennettii*) into two feathers, and the hind wings into three. They
are all of dull colours, brown, grey, or white. The commonest
species is the White Plume Moth (*Pterophorus Pentadactylus*),
which is often to be seen in gardens, or in weedy places. It
expands rather more than an inch. Finally, the little brownish

Twenty-Plume Moth (*Alucita Hexadactyla*), our only representative of the family *Alucitidæ*, has each wing split up into six

Twenty-Plume Moth (*Alucita Hexadactyla*).

feathers.   It is also a common garden insect, and may often be
seen resting with its wings expanded.   It expands rather more
than three-quarters of an inch.

FLOWERS AND INSECTS.

A great many flowers are very attractive to butterflies in the
daytime, and to moths in the evening.   Among those which
deserve special notice are sallow in spring, ivy in autumn,
catch-fly and viper's bugloss on the sea-coast, thistles and rag-
weed in waste places, honeysuckle in hedges, and valerian,
petunia, etc., in gardens.

BOOKS LIKELY TO BE USEFUL TO BEGINNERS.

Kirby's European Butterflies and Moths. Coloured Plates.

Stainton's Manual of British Butterflies and Moths. 2 vols. Woodcuts.

Newman's British Butterflies and Moths. 2 vols. Woodcuts.

Coleman's British Butterflies. Plates.

Wood's Common Objects of the Country. Plates.

Wood's Common British Moths. Plates.

Greene's Insect Hunter's Companion. 1s.

Knaggs' Lepidopterist's Guide.

TABLE OF THE PRINCIPAL GROUPS
OF
BRITISH BUTTERFLIES AND MOTHS.

—◆◆◆—

Rhopalocera.				Butterflies.
NYMPHALIDÆ,—				
Satyrinæ	.	.	.	Brown Butterflies.
Nymphalinæ.	.	.	.	Fritillaries, Tortoiseshells, etc.
ERYCINIDÆ.	.	.	.	Duke of Burgundy Fritillary.
LYCÆNIDÆ.	.	.	.	Hair-Streaks, Blues, and Coppers.
PAPILIONIDÆ,—				
Pierinæ	.	.	.	White and Yellow Butterflies.
Papilioninæ.	.	.	.	Swallow-Tail.
HESPERIIDÆ	.	.	.	Skippers.

Heterocera.				Moths.
SPHINGES,—				
Sphingidæ	.	.	.	Hawk Moths.
Ægeriidæ	.	.	.	Clear-Wings.
Zygænidæ	.	.	.	Burnets and Foresters.
BOMBYCES,—				
Arctiidæ	.	.	.	Tiger Moths.
Lithosiidæ	.	.	.	Footmen.
Liparidæ	.	.	.	Satin Moths.
Psychidæ	.	.	.	Case-bearing Moths.
Notodontidæ.	.	.	.	Puss Moths and Prominents.
Limacodidæ.	.	.	.	
Drepanulidæ	.	.	.	
Saturniidæ	.	.	.	Emperor Moth.
Endromidæ.	.	.	.	Kentish Glory.
Lasiocampidæ	.	.	.	Eggars.
Zeuzeridæ	.	.	.	Goat Moth, etc.
Hepialidæ	.	.	.	Swifts.
NOCTUÆ,—				
Cymatophoridæ	.	.	.	Peach Blossom Moth, etc.
Bryophilidæ.	.	.	.	
Acronyctidæ.	.	.	.	Dagger Moths, etc.
Leucanidæ	.	.	.	Wainscots.
Mamestridæ.	.	.	.	Cabbage Moth, etc.
Caradrinidæ	.	.	.	
Agrotidæ	.	.	.	Yellow Underwings, etc.
Tæniocampidæ	.	.	.	Quakers.
Cosmidæ	.	.	.	
Hadenidæ	.	.	.	

NOCTUÆ (*continued*),—
- *Xylinidæ* . . . . Sharks, etc.
- *Heliothidæ* . . . .
- *Acontidæ* . . . .
- *Erastridæ* . . . .
- *Anthophilidæ* . . . .
- *Brephidæ* . . . .
- *Plusidæ* . . . . Gamma Moth, etc.
- *Amphipyridæ* . . . Copper Underwing and Mouse.
- *Toxocampidæ* . . . Old Lady, etc.
- *Stilbidæ* . . . .
- *Catephidæ* . . . .
- *Catocalidæ* . . . . Red Underwings.
- *Ophiusidæ* . . . .
- *Euclididæ* . . . .
- *Poaphilidæ* . . . .
- *Deltoidæ* . . . . Snout, Fan-Foot, etc.

GEOMETRÆ,— Loopers.
- *Dendrometridæ* . . . Thorns, Emeralds, Waves.
- *Phytometridæ* . . . Carpets and Pugs.

PYRALES,—
- *Pyralidæ* . . . . Meal Moth, Tabby, etc.
- *Botydæ* . . . . . Pearls.
- *Galleridæ* . . . .
- *Phycidæ* . . . . Knot-Horns.
- *Crambidæ* . . . . Grass Moths.

TORTRICES . . . . Bell Moths.

TINEÆ,—
- *Choreutidæ* . . . .
- *Chimabacchidæ* . . .
- *Tineidæ* . . . . Clothes' Moths.
- *Adelidæ* . . . . Long Horns.
- *Hyponomeutidæ* . . . Small Ermines.
- *Plutellidæ* . . . .
- *Gelechidæ* . . . . Flat Bodies, etc
- *Œcophoridæ* . . . .
- *Glyphipterygidæ* . . .
- *Gracilaridæ* . . . .
- *Coleophoridæ* . . . Small Case-Bearers.
- *Lavernidæ* . . . .
- *Elachistidæ* . . . . Grass Miners.
- *Lithocolletidæ* . . . Leaf Miners.
- *Lyonetidæ* . . . .
- *Nepticulidæ* . . . Leaf Miners.
- *Micropterygidæ* . . .

PTEROPHORI . . . . Plume Moths.

ALUCITÆ . . . . . Twenty-Plume Moth.